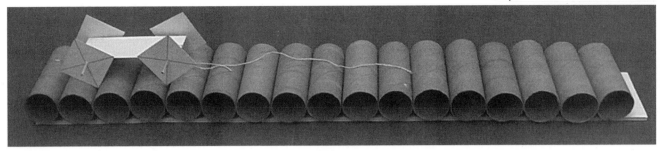

Square Wheels

and Other Easy-to-Build, Hands-On Science Activities

An Exploratorium Science Snackbook

Don Rathjen

Paul Doherty

and the

Exploratorium Teacher Institute

Illustrations by Esther Kutnick

expl◯ratorium   3601 Lyon Street, San Francisco, CA 94123   www.exploratorium.edu

Square Wheels and Other Easy-to-Build, Hands-On Science Activities was developed by Don Rathjen, Paul Doherty, and the Exploratorium Teacher Institute, a part of the Exploratorium Regional Science Resource Center, which is funded by the California Department of Education.

Project Manager: *Kurt Feichtmeir*

Managing Editor: *Pat Murphy*

Developmental Editors: *Judith Brand and Eric Engles*

Designer: *Mark McGowan*

Design and Production: *Carolyn Deacy and Gary Crounse*

Illustrator: *Esther Kutnick*

Photography: *Amy Snyder*

Photo Researcher: *Megan Bury*

Production Editor: *Laura Jacoby*

Proofreader: *Ellyn Hament*

Indexer: *Ty Koontz*

BE CAREFUL! The activities and projects in this work were designed with safety and success in mind. But even the simplest activity or the most common materials could be harmful when mishandled or misused. Use common sense whenever you're exploring or experimenting.

Exploratorium® is a registered trademark and service mark of The Exploratorium.

© 2002 Exploratorium, www.exploratorium.edu

ISBN: 0-943451-55-8

Library of Congress Cataloging-in-Publication Data

Rathjen, Don.
  Square wheels : and other easy-to-build, hands-on science activities / Don Rathjen, Paul Doherty, and the Exploratorium Teacher Institute illustrations by Esther Kutnick.
    p. cm. — (An exploratorium science snackbook series)
Includes bibliographical references and index.
  ISBN 0-943451-55-8
  1. Science—Experiments. I. Doherty, Paul. II. Kutnick, Esther, ill. III. Exploratorium Teacher Institute (San Francisco, Calif.) IV. Title. V. Series.
  Q164 .R38 2002
  507'.8—dc21
                        2001008355

Printed in the United States of America

GE® Party Bulb is a registered trademark of General Electric Company.

Glow Max® phosphor-coated, glow-in-the-dark paper is a registered trademark of the Riverside Paper Company.

Golden® Acrylic Phosphorescent Green Glow in the Dark Acrylic Paint is a registered trademark of Golden Artist Color, Inc.

Hopper Popper® is a registered trademark of Dynatoy International Inc.

Liquitex® is a registered trademark of ColArt Americas.

Mylar® is a registered trademark of E.I. du Pont de Nemours and Company.

Nerf® is a registered trademark of the Tonka Corporation.

Phillips® is a registered trademark of Facom.

RadioShack® is a registered trademark of the RadioShack Corporation.

Rebound® is a registered trademark of The Exploratorium.

String Ray® is a registered trademark of With Design in Mind.

Super Ball® is a registered trademark of Wham-o.

Teflon® is a registered trademark of E.I. du Pont de Nemours and Company.

Testors® Gloss Enamel is a registered trademark of The Testor Corporation, an RPM Company.

Wellington® Braided Nylon Chalk and Mason Line is a registered trademark of Wellington Inc.

X-Acto® knife is a registered trademark of Hunt Manufacturing Co.

Image Credits

All illustrations by Esther Kutnick and all photos by Amy Snyder unless otherwise noted. Page 3: J. R. Eyerman/TimePix. Page 8: Randy Comer © 2000 Exploratorium, www.exploratorium.edu. Pages 9, 10, 11, 17, 18 (top), 19 (top), 20 (left), 26, 29, 30 (bottom), 31, 33 (top and middle), 41, 43, 47 (bottom), 48 (right), 49, 50, 51, 52, 53 (top), 56, 64, 65, 66 (left), 74, 81, 82, 85, 86, 89, 90, 105, 106, 113, 114, 115 (top and bottom), 120 (top and middle), 121, 126 (top), 129, 130: Don Rathjen. Page 15: courtesy of A. Geim, University of Manchester, UK. Page 18: (bottom) courtesy of Massimo Borei, http://www.funsci.com. Page 20: (right) Edgar Fahs Smith Collection, University of Pennsylvania Library. Page 23: (left) © John S. Shelton; (right) courtesy of Benoit Mandelbrot; (bottom) courtesy of Jim "jimonade" Harney. Page 27: (right) courtesy of Dr. Hubert Dolezal. Page 32: © CORBIS. Page 33: (bottom) courtesy of www.oldtemecula.com. Page 37: (bottom) courtesy of Paul Doherty. Page 48 (left) and 61: © 1984 Facts on File, New York, NY. Page 54: courtesy of U. S. Department of the Navy, Naval Historical Center, Washington, DC. Page 55: courtesy of Pat Murphy. Page 66: © Bettmann/CORBIS. Page 72: (left) Mauritshuis, The Hague; (right) courtesy of Laura Jacoby. Page 75: (top) The Metropolitan Museum of Art, Harris Brisbane Dick Fund, 1941 (41.48.3). Page 80: courtesy of Robert Greenler, from *Rainbows, Halos and Glories*, Blue Sky Associates, www.blueskyassociates.com. Page 83: (right) courtesy of Richard I. Gregory, © 1997 Princeton University Press, Princeton, NJ. Page 87: (top) © 1998 David Farley, http://sunsite.unc.edu/Dave/drfun.htm; (bottom) Collection Fons Vanden Berghen, Hale, Belgium. Page 99: © Harold & Esther Edgerton Foundation, 2001, courtesy of Palm Press, Inc. Page 107: (top) Marni Fylling; (bottom) Illustrations of Surgical Instruments of Superior Quality, 1915, NY. Page 111: courtesy of Gene Easter and Bill Reitz. Page 115: (middle right) © Greg Helgeson. Page 117: courtesy of Paul Hewitt. Page 122: © Jack Fields/CORBIS. Page 128: © E. O. Hoppe/CORBIS. Page 131: (bottom) National Gallery of Art, Washington, DC. Collection of Mr. and Mrs. Paul Mellon.

Contents

Introduction

Square Wheels and The Exploratorium Science Snackbook Series

Welcome to *Square Wheels and Other Easy-to-Build, Hands-On Science Activities,* the second book in the Exploratorium Science Snackbook Series. This book contains 31 new science "snacks"—simple, inexpensive activities, demonstrations, and exhibits designed to provide you with the opportunity to conduct your own inquiries into the phenomena of science. *Square Wheels* is the successor to the original *Exploratorium Science Snackbook,* published in 1991, which contained 107 snacks based on exhibits at the Exploratorium. Both the original *Snackbook* and *Square Wheels* were developed by teachers for teachers, students, and anyone who wants to learn science by doing science. In *Square Wheels,* about half the snacks are based on Exploratorium exhibits; the rest were developed by staff and teachers at the Exploratorium Teacher Institute.

Students at the Exploratorium use a mirror with horizontal gaps in it to see a face that is partly theirs and partly a friend's in a snack called "Your Father's Nose."

The Exploratorium

The Exploratorium is a hands-on museum of science, art, and human perception in San Francisco. It's been called a scientific funhouse, a giant experimental laboratory, even a mad scientist's penny arcade.

Each year, more than half a million visitors come to the Exploratorium; 60,000 of them are students on field trips. We often see our visitors become science teachers: They discover things for themselves, then show their discoveries to someone else. Kids turn to their parents and say, "Look at this!" Visitors of every age have the opportunity to use more than 600 hands-on, interactive exhibits to discover for themselves that science is fun.

The Exploratorium presently supports two teacher-training programs: the Institute for Inquiry, for elementary school staff development leaders, and the Teacher Institute, for middle and high school science and math teachers. These programs teach science to teachers using hands-on discovery—the same method we encourage them to use in teaching science to their students.

How the Snackbook Series Began

Ever since the Exploratorium opened in 1969, teachers from around the San Francisco Bay Area have brought their classes here to get their kids excited about science. And from the very beginning, teachers have asked, "How can I bring these exhibits home to my classroom?"

This was a challenge the Exploratorium couldn't ignore. We already had three volumes of the *Exploratorium Cookbook: A Construction Manual for Exploratorium Exhibits,* which were written to help other museums create duplicates of Exploratorium exhibits. However, *Cookbook* instructions were complex and demanding, and relied on materials and skills well beyond those available to the average teacher. The Teacher Institute helped a group of teachers to write the book they wanted—a book telling how to build simple, inexpensive, classroom-sized versions of Exploratorium exhibits.

For over a decade, more than two hundred teachers and Exploratorium staff members have been creating and testing recipes for classroom science activities and exhibits. Time after time, the teachers experienced the joy of discovering new ways to do science. The excitement of the *Snackbook* brainstorming sessions was contagious. Teachers told us that they felt a rejuvenated interest in teaching science. With assistance from the Exploratorium's own science, writing, and graphics staff, the initial recipes became the first *Exploratorium Science Snackbook,* or the *Snackbook,* for short. *Square Wheels* is the second volume in The Exploratorium Science Snackbook series.

Each snack was developed by one or more teachers. Sometimes their innovations even improved on the original museum exhibit. For example, the "Circuit Workbench" snack in *Square Wheels* allows for more explorations than the Exploratorium exhibit. In *Square Wheels,* teachers have also created some snacks that are not based directly on Exploratorium exhibits, but which have the same inspiration of playful exploration as the exhibits.

What's in a Snack?

The design of *Square Wheels* reflects the needs and requests of the teachers who created it. The snacks are divided into easy-to-follow sections that include instructions, advice, and helpful hints.

- Each snack begins with a photograph of the finished exhibit.

- There's a list of the materials needed, and suggestions on how to find them.

- Other sections give complete assembly instructions and contain descriptions of how to use the completed experiments.

- Because the teachers insisted that correct scientific explanations accompany the hands-on activities, each snack explains the science behind the phenomenon being demonstrated.

- So What? and Did You Know? sections contain interesting bits of additional historical and scientific information.

- Going Further sections suggest ways to begin your own explorations.

How to Use the Snackbook

After its publication in August 1991, the original *Exploratorium Science Snackbook* rapidly found its way far beyond the San Francisco Bay Area. Within a week, for example, the Exploratorium had received a request from the Australian outback asking us to help teachers there find a supplier of plastic mirrors for their *Snackbook* experiments.

We also discovered that a wider range of teachers was using the *Snackbook* than we had originally expected. Though the *Snackbook* was written primarily for high school teachers, we began to hear of successful applications in elementary schools, middle schools, colleges, and universities. We also heard from local science teachers who had special education classes or were working with students learning English as a second language. While these students had great trouble learning science from their textbooks, many excelled at building snacks and investigating science.

We found that the *Snackbook* was particularly useful in school districts where science department funding was tiny. With its emphasis on inexpensive or scrounged parts, the *Snackbook* gave teachers in less-well-funded districts a way to do hands-on science activities on a tight budget. Teachers who had never been to the Exploratorium asked us how to weave the hands-on activities described in the *Snackbook* into their classroom lessons.

There are many ways to incorporate interactive science activities into your classroom. In the section that follows, we will show you how teachers use the *Snackbook* to help their students create science exhibits of their own. We hope that this book will give you a few new ideas on how to use hands-on science in your classroom.

Teachers' Tips on Using Snacks

Teachers from around the globe have found snacks to be natural teaching tools, highly adaptable to the individual needs of their classrooms. Exploratorium snacks have been used in a wide range of settings, including traditional science, special education, and English as a second language classes to teach scientific concepts to a wide range of students, including at-risk, underserved, and college-level teacher-preparation students

Putting together a snack can be a compact science lesson in itself, and all teachers know that science becomes "real" when students put something together and see for themselves how the process works. Following are several accounts from teachers who have created "mini-Exploratoriums" in their own classrooms, tailoring the snack process to suit their specific needs.

Snacks as an Alternative to a Traditional School Science Fair

"As part of the curriculum, students present a mini-Exploratorium at our Middle School Science Night. Students select their projects from an Exploratorium activity publication, or from snacks on the Exploratorium Web site, which is available on the computers in our school library. Students are made aware of the expectations and requirements of the assignment, are given a printed sheet that outlines the project for them, and are shown some of the past year's exhibits in class for ideas. The mini-Exploratorium projects are intended to be interactive and relatively easily constructed with a minimum of time and expense. Parents are encouraged to be supportive, but to keep their role to an advisory one.

"Each snack is to be accompanied by a self-standing display board that provides the necessary information for a person to interact independently with the activity. At the very least, it should include the title, the materials needed to assemble the snack, the To Do and Notice, and the What's Going On? Any additional information about the snack that provides some insight for the participant is encouraged.

"Since this assignment is one that is completed independently and mostly at home, it is important that students have a definite plan for its completion. Students can ask for advice if they need help. Grades for the mini-Exploratorium snacks are based on a number of considerations, some of which are listed below.

- The degree of challenge the snack presents

- Innovation

- Classroom presentation

- Science night presentation

- Effectiveness of the exhibit and display board

- Meeting all due dates

"The criteria are mostly subjective and are intended to be. This assignment is not like taking a test—effort, ingenuity, and creativity are as much a part of this assignment as the completion of the display. If students challenge themselves and put in the necessary effort, they will do fine. The whole experience has proven to be a very positive experience for all involved."

—Eric Kielich, Science Department Head and Teacher, Mount Tamalpais School, Mill Valley, Calif.

Snacks on Tour

"Students individually or in small groups select snacks from the *Exploratorium Science Snackbook* or other sources, gather materials, construct the snacks, learn a bit about the science involved, and make graphics explaining how to use the snacks and how they work. Then a time is set for the big event, our classroom is transformed into an exhibition hall, and guests are invited to peruse the snacks on display, with students as explainers.

"The guests are the key element. They may be other students, teachers, parents, and staff from the school, the school board, local officials, local press, or others. When students get the opportunity to create work for an audience beyond the teacher and the classroom walls, the results are truly staggering. The satisfaction and confidence boost that students gain may even outshine the enormous amount of self-led learning that goes on.

"Another dimension is added by going 'on tour.' Instead of making arrangements for guests to come to us, we take our 'mini-Exploratorium' to them. Our guests are primarily students at other schools. We go to at least one school and try to hit two or three classes at minimum. We have shown our exhibition to 500 students in three other schools on a couple of occasions.

"However, when the introductions have been done, and the small groups of guests are moving from snack to snack, there is nothing to do but sit back, smile, and take some pictures. While it is advisable to have a small tool kit (hot glue, tape, wire, pliers, scissors, spare parts, etc.) on hand for emergency breakdowns, these presentations are often classes that run themselves.

"This kind of 'exhibit on tour' is a win-win-win-win situation. Both sets of students gain enormously, the teacher to whose class you are presenting sees that this is doable and may be inspired to do it in the future, and you have provided exceptional quality education while also getting considerable satisfaction.

"The logistics of the mini-Exploratorium can be frightening: finding materials, coordinating students and guests, ensuring correct understanding of concepts with each team, and making sure everyone is done on time. But, taken a step at a time, it is manageable, and the results are well worth the effort.

"The mini-Exploratorium can be a great way to culminate the year or an exciting ice-breaking project with a new class to start the year off right. It constitutes a fantastic science unit for students from elementary through high school and can be focused on any area of science. Of course, a great way to begin or end this unit is with a field trip to your local science museum."
—Curt Gabrielson, former Teacher Institute staff

Interactive Science Museum

"About ten years ago, using the *Exploratorium Science Snackbook* as a resource, I built twenty exhibits and set up a museum. Since durability was a concern, I beefed up each 'snack' whenever possible. My water spinner with its acrylic tank and $\frac{3}{4}$-inch plywood turntable has survived for twelve years. I used the write-ups in the *Snackbook* to create exhibit signs. The *Snackbook* provided a template for me to use when I created my own snacks. The museum currently has about forty exhibits.

"Building the museum has not been easy. Most of my projects are built with scrap materials so the size and shape depend on what is in my scrap woodpile. I collect and save lots of junk. I purposely keep the projects simple because I believe that a slick production sometimes hides the science. Whenever possible, I try to keep the projects small enough so that they will fit into those red flip-top storage crates that are sold at Home Depot. In each case, I store the exhibit and the sign explaining it.

"After a time, it can become difficult to maintain so many exhibits. I have had to let some exhibits go, and I have replaced some and added others. I have several groups that have reimbursed me for expenses or have paid to have me put on a science night for their school. These events energize me and help keep the museum going. I have set up the museum at California State University Hayward, the lobby of The Tech Museum in San Jose, and at the Union City Science, Earth, and Health Festival. I have also set up science nights for local schools. Often, my students will serve as docents.

"The *Snackbook* has been valuable resource. As I built 'snacks,' I began to think more creatively, developing a taste for what motivates and inspires. I like to think of myself as a snack chef!"
—Charles Reynes, Science Teacher,
Creekside Middle School, Castro Valley, Calif.

Putting Exploratorium Snacks to Work for You

Here are more teacher tips that will help enhance the effective use of the *Square Wheels* snackbook and make the Exploratorium's "hands-on, minds-on" experience an integral part of your science teaching plan.

Ways to Use Snacks

- High school students or middle school students can use snacks as part of a science show for students in lower grades. In reality, students of all ages can present snacks to younger, same grade, or even older students.

- Snacks can be the basis of enrichment projects for interested students.

- Students may build snacks as individual or small group science projects, either as an assignment or for extra credit.

- Snacks may be used for an interactive science museum at the school or district level.

- Teachers can build snacks for demonstration or interactive exhibit use.

- Snacks can be used as the basis for a "mini-Exploratorium," which can serve as an alternative to a traditional science fair and can be shared with the entire school and beyond.

- Students can use snacks as a science show for their community, e.g., at parents night, in a convalescent home or church, in a mall booth, or for service clubs. Go on tour!

- Snacks can be used as labs or lab supplements.

Real Teachers' Real Tips

- Keep a stock of common snack supplies in class.

- Stock simple tools in the classroom for students to use.

- Make finding materials fun (e.g., assign bonus points for bringing items in, organize a swap meet or a treasure hunt).

- Find outside sources for inexpensive materials, such as flea markets, thrift shops, recycling centers, garage sales, etc.

- Know where students can obtain less-accessible materials.

- Modify and improve snack designs yourself (build a better mousetrap; a modified snack may often be an improvement over the original). Encourage your students to do the same.

- Encourage improvisation and innovation—have students design their own snacks and devise their own experiments.

- Have snack-building teams help each other (peer review).

- Keep two copies of *Square Wheels* in your classroom—one can be checked out overnight, the other is for class use only.

- Fill your classroom with attention-getting, even distracting, science materials for students to explore on their own.

- Have students present oral and written reports.

- Use *Square Wheels* for classroom reference; photocopy actual snacks for student use once they have been chosen or assigned.

- Allow enough time for students to gather the materials, assemble, tinker, and perfect their snacks based on their explorations.

- Remind students that testing a snack to make sure it works is part of the process, and that things may not work perfectly the first time they are used. Reinforce that science is trial and error—it's OK, even desirable, to make mistakes.

- Build snacks and other projects yourself. Be flexible: Admit that you don't have all the answers and that your attempts may sometimes fail, too. Share the process with your class, and work with them to figure out what went astray with your project.

- Consider your first use of snacks as an experiment in itself; keep track of how you handled things so that you can make modifications in the process as necessary the second time around. It may take you a couple of years to find the mode of use that works best for your situation.

How to Get Cooking with the Square Wheels Snackbook

In the upper left-hand corner of each snack, you'll find a Topic Guide for that snack. The topics covered in *Square Wheels* are:

Chemistry	Fluids	Mathematics
Color	Heat	Mechanics
Earth Science	Inquiry	Perception
Electricity	Light	Sound
Energy	Magnetism	Waves

To find a complete list of snacks related to any particular topic, you can refer to the Topic Index at the back of *Square Wheels,* where snacks are listed in topic order.

Each snack has also been assigned a Challenge Meter rating. You'll find the Challenge Meter below the snack title, at the upper left corner of the opening photo. To calculate the degree of challenge for a snack, we considered several factors: how easy or difficult it was to find the necessary materials; how complicated the assembly process was; and how much time was needed to complete the snack. A more complete understanding of each snack's ease or complexity can be gained by reading the Materials list and the Assembly directions.

Materials are listed in order of use, with a few exceptions when we have grouped like items, such as drill bits of different sizes.

In some of the snacks, we have included Alternative Construction tips, which suggest time-saving strategies, alternate construction materials, or variations on the snack. Occasionally, you will see Helpful Hints, which contain extra information to help with tricky aspects of construction or offer more background on technical aspects of the snack.

3-D Shadows

Are they just an illusion?

All shadows are two-dimensional—that is, flat. But you can fool your brain into seeing a three-dimensional shadow with this snack. First, you make an object cast two separate shadows. Then, with the appropriate filters over your eyes, these two shadows appear as one shadow with three-dimensional depth.

Materials

- thread, about 20 in (50 cm)
- fork (any material except clear plastic)
- tape (cellophane or masking)
- white poster board or white paper large enough to cover most of the bottom of the box

- cardboard box approximately 12 in × 15 in × 10 in (31 cm × 38 cm × 25 cm), no lid or top
- nail or ice pick, for poking a small hole through the side of the box
- 1 red light bulb (e.g., GE Party Bulb)
- 1 green light bulb (e.g., GE Party Bulb)

- 2 plug-in lamp-base sockets
- outlet strip (and extension cord, if necessary)
- red and green filters (gels, plastic, cellophane, or transparencies); available at theatrical supply stores, plastics stores, or office supply stores (report covers)

1 Tie the thread around the handle of the fork in a tight loop. Adjust the location of the loop by sliding it along the handle so that when the fork is hanging on the end of the thread, the fork balances horizontally with its prongs aimed slightly upward. When you have the fork properly balanced, tape the loop in place with a small piece of tape.

2 Tape the paper onto the bottom of the box to create a white projection screen.

3 Place the box on a table with its opening facing you and with its long side on the table.

4 Use the nail or ice pick (or other similar implement) to poke a hole in the middle of what is now the top side of the box. Stick the loose end of the thread that is tied to the fork up through the hole from the inside of the box. Pull the thread up until the fork hangs about halfway down inside the box. Keep the fork suspended at this height by taping the thread to the top of the box. The fork should have enough room to rotate freely on the end of the thread without hitting the back of the box. (If it doesn't, relocate the hole so that it does.)

5 Screw the red and green bulbs into the sockets, and plug the sockets into the outlet strip so that there is one empty outlet between them.

6 Place the outlet strip parallel to the bottom edge of the box, and about three times as far away from the screen as the fork (that is, about half of the depth of the box out from the edge of the box). Position the outlet strip so that the bulbs are both approximately the same distance from the fork.

7 Plug in the outlet strip and turn on the switch to light the bulbs. When the full length of the fork is hanging straight across the box, the shadows it casts should fill about half to three-quarters of the width of the screen; if they don't, adjust the position of the power strip a little so that they do.

Alternative Construction

In the assembly, the shadows are projected on the front of the screen, and you can see the actual fork at the same time you are looking at the shadows. With the same or similar materials, however, you can build a 3-D shadow projector that uses rear projection and hides the lights and the actual fork from view. The easiest way to do this is to cut a large rectangular hole in the bottom of the box before you tape the white paper to the bottom, and to set up the fork and lights as described. The shadows cast on the paper screen will be visible through the paper from the back side. You can also improvise a way of making a translucent screen (with white paper or a piece of thin white cloth) and hanging the fork so that the lights cast shadows of the fork on the screen. Then—with the room lights sufficiently dimmed—you can view the screen from the other side to see the three-dimensional image of the fork.

To Do and Notice

Dim the room lights a bit if they are bright. Tap the hanging fork so that it rotates a little.

Notice that one shadow is red and the other green, and that they are slightly separated.

Hold a red filter over one eye and a green filter over the other eye, and look at the shadows. You should now see only one shadow, and it should appear black and three-dimensional. Notice that this shadow seems to be either behind the actual fork or in front of it, and that when the fork rotates, the shadow rotates as well.

Switch filters to the opposite eyes. Notice that the shadow changes its apparent location—if it seemed to be behind the actual fork before, it now seems to be in front of it, and vice versa. Also notice that the direction in which the shadow rotates may appear to have changed.

➡ Helpful Hint

About 10 percent to 15 percent of people have difficulty with stereo vision and may not be able to perceive the illusion created by this snack.

What's Going On?

Most of the white screen is hit by both red light and green light, which together appear yellow. But there is an area of the screen where the fork blocks the red light, and another area where the fork blocks the green light. Where the fork blocks the red light, only green light hits the screen, and a green shadow of the fork appears. Likewise, where the fork blocks the green light, only red light hits the screen, forming a red shadow of the fork. Because the sources of the two colors of light are at different locations, the red and green shadows are formed at slightly different places on the screen.

Each filter has the effect of making its same-color shadow blend into the background, and therefore seem to disappear, while making the different-color shadow appear black. The red filter lets the red light from the red and green background (seen as yellow) and the red light from the red shadow reach your eye. As a result, the red shadow and the background both look red, so the red shadow itself can no longer be distinguished. But only green light comes from the green shadow. Because green light is absorbed by the red filter, it's prevented from reaching your eye, making the green shadow appear black. In a similar fashion, the green filter over your other eye causes that eye to see the red shadow as black and prevents it from seeing the green shadow.

Each eye, therefore, sees a black shadow in a slightly different place on the screen. This is just like what occurs in ordinary vision: Each eye sees a slightly different view of whatever you are looking at, since your eyes are at two different locations on your head. Just as your brain does with the two views it gets of objects in everyday vision, your brain fuses the two shadows, interpreting the differences between them as clues to depth. As a result, you see a single three-dimensional shadow, just as you see single three-dimensional views of ordinary objects.

This audience is watching Bwana Devil, the first full-length color 3-D motion picture.

Reversing the positions of the colored filters is equivalent to reversing the views that your eyes get. In many cases, but not all, this reversal causes your brain to reverse its depth perception: Objects that appeared far away before will now appear to be near, and vice versa.

The reversal of the direction of rotation is an extension of the depth-perception reversal, but the actual direction of rotation perceived by a given observer can be very subjective.

So What?

Three-dimensional images may be shown on a flat computer screen or printed on a page of paper by printing one eye's view in red and the other eye's view in green. The viewer looks at the images through the red and green filters used in this snack. Each eye sees its color-coded image, and the brain assembles the different images into a 3-D view. This technique, known as red-green anaglyph photography, was used to send three-dimensional images of Mars over the Web during the Pathfinder mission.

Did You Know?

Two Different Perspectives

To see how each of your eyes gets a different view of the world, try this: Hold up one of your fingers at arm's length from your face, but focus your attention on a more distant object behind it, perhaps something on a wall across the room. Close one eye and note the position of your finger relative to the distant object, then open that eye and close the other eye. You should see the position of your finger shift horizontally, to the left or right, relative to the distant object. Careful observation will show you that each eye is getting a slightly different view of your finger.

Going Further

3-D Glasses

Try making a pair of red-green viewing glasses that you can wear instead of holding the colored filters in front of your eyes. Try to design the glasses so that you can easily reverse the color that each eye looks through.

Other Shadows

Make objects out of wire, straws, balsa wood, and so on to replace the fork. A small hollow cube is traditionally used in shadow-projection experiments.

3-D Photos

Try using the red and green filters to view 3-D photos of the Exploratorium building on the Web at: www.exploratorium.edu/history/PPIE-3D/index.html.

Credits & References

This snack is based on the Exploratorium exhibit of the same name.

Falk, David, Dieter Brill, and David Stork. *Seeing the Light: Optics in Nature, Photography, Color, Vision, and Holography.* New York: John Wiley & Sons,1986.

Gregory, Richard L. *Eye and Brain: The Psychology of Seeing,* 4th ed. Princeton, N.J.: Princeton University Press, 1990. See page 75 for a brief discussion of reversal of perceived depth due to the right eye receiving the left eye's view, and vice versa.

Gregory, Richard L. *The Intelligent Eye.* New York: McGraw Hill, 1970. See Chapter 3, titled "Ambiguous, Paradoxical and Uncertain Figures."

Bits and Bytes

Record a binary message on the world's cheapest digital tape recorder.

When you listen to a CD, use a computer, or send a message to a pager, you are dealing in digital information, which is represented by nothing more than the binary digits 1 and 0. In this snack, you can create your own digital message by using a magnet to encode 1s and 0s in small strips of tape from an audio cassette. You can then "read" a partner's encoded message with the same magnet.

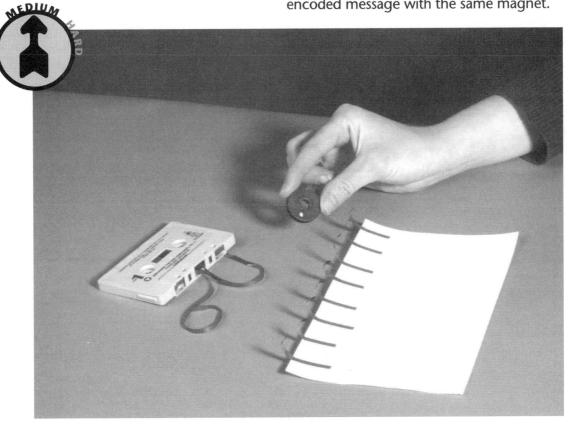

Materials

- old audio cassette tape (one you can destroy)
- scissors
- transparent or translucent tape
- 4 index cards, 5 in × 8 in (13 cm × 20 cm) or larger

- magnetic compass
- 1 ceramic "donut" disk magnet, approximately 1 in (2.5 cm) in diameter (e.g., RadioShack #54-1888, pack of five); alternatively, a hard ceramic refrigerator magnet

- white correction fluid or masking tape
- pen or pencil
- a partner

You and your partner should each make a set of cards and encode them as described below. Then you should trade cards and decode them. (If necessary, you can do the snack yourself, but you will have to decode your own cards.)

1 Pull 6 feet (about 2 m) of tape out of the cassette and snip it off with the scissors. From this long piece of tape, cut 32 short strips of tape, each about 2 inches (5 cm) long.

2 Notice the curl in the short pieces of audiotape. Tape one of the pieces of audiotape, with the curl up, to one of the index cards in the position shown in figure 1 (about a half inch [1.25 cm] from the end of the card). The piece of tape should curl up beyond the edge of the card.

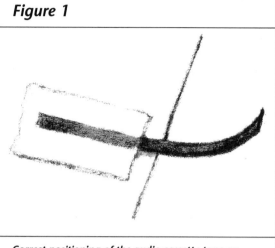

Figure 1

Correct positioning of the audio cassette tape on an index card

3 Tape seven more of the short strips to the edge of the card in the same way, each about 1 inch (2.5 cm) away from the previous strip.

4 Tape strips of audiotape to the other cards in the same way to create four identical cards. Number the cards 1 through 4.

5 Place the magnetic compass on a table and note which end of the compass needle points north and which end points south.

6 Find the face of the magnet that attracts the end of the compass needle that points south and repels the end of the compass needle that points north. This face is the north pole of the magnet.

7 Mark the north pole of the magnet with a dot of correction fluid (or a piece of masking tape).

To Do and Notice

Place the north pole of the magnet on the attached end of one of the audiotape strips. Gently slide the magnet along the strip from the attached end to the free end, as shown in figure 2. One "stroke" is enough. The magnetic field of the magnet is strong enough to magnetize the tape.

Like any magnet, the magnetized strip of audiotape now has a north pole and a south pole. Is the free end of the strip a north pole or a south pole? To find out, bring the north pole of the magnet back toward the magnetized tape slowly as shown in figure 3—but don't let the magnet actually touch the tape or you'll reverse its polarity. Does the magnet attract the end of the tape or repel it?

The north pole of the magnet should attract the free end of the tape strip. This tells you that the free end is a south magnetic pole (remember that different poles attract). Flip the magnet around and bring its south pole end toward the magnetized tape slowly. Notice that the magnet now repels the tape.

As a general rule, the magnetic polarity of the free end of the strip of

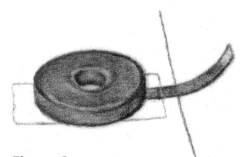

Figure 2 *Slide the magnet along the tape.*

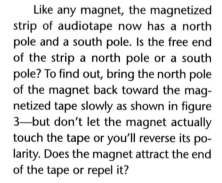

North pole

Figure 3 *Bring the north pole of the magnet slowly toward the end of the tape.*

tape will be opposite that of the magnetic pole you use to magnetize the strip. In other words, if you rub the north pole of a magnet along the strip toward the free end, you create a south magnetic pole at the free end of the strip, and if you rub the south pole of a magnet along the strip toward the free end, you create a north magnetic pole at the free end of the strip.

Now that you know how to magnetize the tape strips in two different ways, you can create a binary message for another person to read. For this purpose, let's say that a strip with a south pole at its free end (one that is attracted to a north magnetic pole) stands for the number 1, and that a strip with a north pole at its free end (one that is repelled by a north magnetic pole) stands for the number 0. To summarize:

1 = strip attracted by a north magnetic pole

0 = strip repelled by a north magnetic pole

Look at the table of binary codes. Notice that there is a unique 8-digit series of 0s and 1s for each letter of the alphabet (these are the same codes that computers use; each capital letter, each number, and each punctuation mark also has its own unique binary code).

Pick a common four-letter word to put into binary code. Each of your four cards will represent one letter of the word.

Put the first letter of your word into binary code by magnetizing the eight strips of card 1 so that they correspond to the binary code for that letter. Start with the tape strip that's on the left when you view the card with the strips facing toward you. After you've coded the first letter of your word, code the next four by repeating the process on cards 2 through 4.

When you have completed magnetizing your four cards, trade cards with your partner. Read the binary code of each card by holding the north pole of the magnet near each

Binary Codes

Letter	Code
a	01100001
b	01100010
c	01100011
d	01100100
e	01100101
f	01100110
g	01100111
h	01101000
i	01101001
j	01101010
k	01101011
l	01101100
m	01101101
n	01101110
o	01101111
p	01110000
q	01110001
r	01110010
s	01110011
t	01110100
u	01110101
v	01110110
w	01110111
x	01111000
y	01111001
z	01111010

strip but not touching it. (Remember that a strip attracted to the north pole of the magnet stands for a 1 and that a strip repelled by the north pole stands for a 0.) Figure 4 shows a card corresponding to the letter *a* (01100001).

What's Going On?

The cassette tape is covered with a layer of magnetizable particles, commonly oxides of iron. These particles become magnetized when they are exposed to a strong magnetic field from a nearby magnet. They keep this

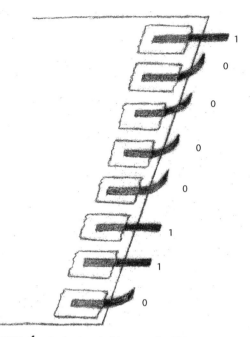

Figure 4 *The letter* a *in binary code. This illustration shows the curvature of each tape when the magnet is held near it. When the magnet isn't nearby, all of the tapes will probably have about the same curvature.*

magnetization unless they are remagnetized by another magnetic field. In this sense, the tape has a magnetic memory, which lets you create two types of tape strips—those with north magnetic poles at their free ends, and those with south magnetic poles at their free ends.

Since each tape strip stands for one binary digit, it represents what is called a *bit* of information. A bit is the smallest unit of binary information. Each card, with its eight bits of information, represents what is called one *byte* of information.

So What?

This snack mimics the way that computers store and retrieve information on floppy disks and hard drives, both of which are coated with magnetic material. Information is stored in the form of tiny areas of attraction and repulsion that stand for the binary digits

1 and 0. All programs and data are in the form of binary numbers. Very small magnetic *write heads* and *read heads* inside a disk drive put the information on the disk and retrieve it.

One gigabyte (GB) is equal to 1000 megabytes, or one billion bytes. Because each card in this snack holds one byte of information, you'd need a billion cards to hold one GB.

Did You Know?

How Many Bits?

To assign a unique binary number to each letter of the alphabet you would only need 5-bit numbers. To encode capital letters, numbers, and symbols as well, you need at least 7 bits. The ASCII code (American Standard Code for Information Interchange) uses 7 bits, and the EBCDIC code (Extended Binary Coded Decimal Interchange Code) uses 8 bits. These codes are used to transfer information and represent data in digital systems and computers.

As Fast as Lightning

At present, a digital cassette recorder will record at least 44,000 16-bit binary numbers a second in order to accurately record music. How long do you think it would take you to "record" one 16-bit binary word using your strips of audiotape and a magnet?

Digital Versus Analog

Digital technology is based on the simple "north-south" or "on-off" quantities of binary numbers. Analog technology, in contrast, is based on continuously changing values.

In general, digital operations are easier to accomplish than analog operations, and for this reason most modern electronic devices are based on digital technology.

How Hungry Are You?

Since an 8-bit binary number is called a *byte,* information scientists have dubbed a 4-bit number a *nibble!*

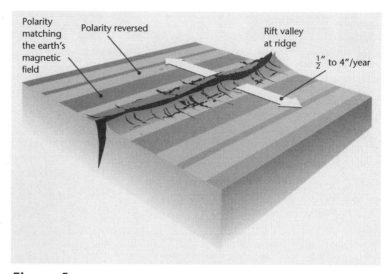

Figure 5 *The spreading ocean floor records the magnetic field of the earth.*

Going Further

By the Numbers

Each of your 8-bit binary codes can represent a number. If there is a 1 in a place, then that place represents that power of 2. If there is a 0 in that place, then that place represents 0. The number is obtained by adding together all the values represented by 1s. For example, the 8-bit binary number 00101011 represents the number 43, the sum of the values of places with 1s:

0	0	1	0	1	0	1	1
		2^5		2^3		2^1	2^0
		32		8		2	1

$$2^5 + 2^3 + 2^1 + 2^0 = 32 + 8 + 2 + 1 = 43$$

Try determining the numerical values of the 8-bit binary numbers you used for your four-letter word.

What's Next?

Study the table of binary codes for the 26 letters of the alphabet on page 7. Try to discover the pattern. Then explain the pattern and predict the next codes in the sequence.

Magnets on the Sea Floor

The earth's magnetic field flips its direction from time to time. (The last time it flipped was about 780,000 years ago.) These flips of polarity are marked in magnetizable rocks formed at the mid-ocean rifts. The rocks well up as molten material; as they cool, they capture the magnetic polarity of the earth at that time. The sea-floor spreading associated with plate tectonics and continental drift then moves the solid rocks outward in both directions from the mid-ocean ridges, forming alternating bands of north- and south-magnetized rock (see figure 5).

Geophysical research ships can sail over the mid-ocean ridges and measure the magnetization of the rocks below to learn about the rate of sea-floor spreading. Try to model this process with your tape strips.

Credits & References

This snack is based on activities developed by Linda Shore and Paul Doherty of the Exploratorium Teacher Institute.

Bloomfield, Louis. *How Things Work: The Physics of Everyday Life.* New York: John Wiley & Sons, 1997. See the sections on Digital Recording (pp. 581–582), Representing Numbers (pp. 477–478), and Storing Numbers and Information (pp. 478–480).

Circuit Workbench

Circuits so interesting you can't possibly get board.

The simple circuit board you build in this snack allows you to easily connect small holiday lights in a variety of ways. By observing the effects of connecting the lights in different ways, you can learn some of the characteristics of series and parallel circuits.

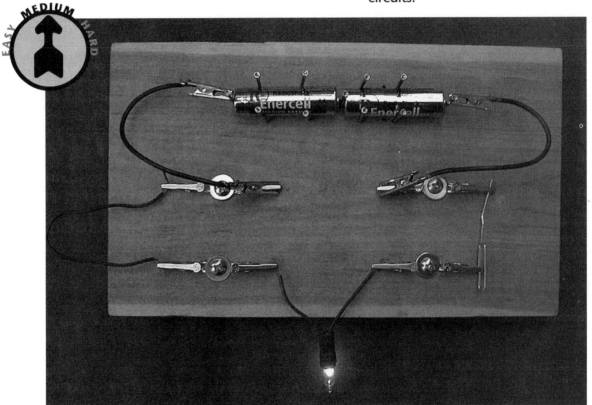

Materials

- hammer
- 11 finishing nails, $1\frac{1}{2}$ in
- wooden board (ordinary 1 in $\times$ 6 in pine shelving), about $5\frac{1}{2}$ in $\times$ 9 in (14 cm $\times$ 23 cm)
- needle-nose pliers
- 12 mini alligator clips (RadioShack #270-380A, pack of 12)

- 4 flat washers (SAE 10)
- 4 sheet-metal screws (#8, $\frac{5}{8}$ in, Phillips pan head)
- Phillips screwdriver
- 2 AA batteries
- short string of miniature Christmas-tree lights

- wire stripper
- 3 pieces of wire, each about 6 in (15 cm) long; you can use extra pieces from the light-bulb string, or #20 or #22 solid or stranded wire
- several metal paper clips

1 Use the hammer and one of the nails to make small pilot holes for screws in the board, at the approximate locations of the four screws in figure 1 (the exact position of the screws is not crucial).

Figure 1

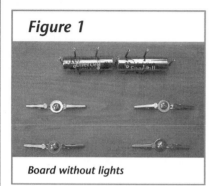

Board without lights

2 Use the needle-nose pliers to bend the two small tabs on the ends of eight of the alligator clips outward, so that the entire end of each clip is flattened as shown in figure 2.

Figure 2

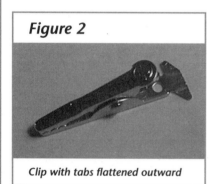

Clip with tabs flattened outward

3 Put a washer on each of the screws, and screw the screws about halfway into the pilot holes.

4 Put the flattened ends of two alligator clips under one of the washers. Position the clips so that they point in opposite directions, parallel to the length of the board, as shown in figure 1. Tighten all the screws until the clips are held firmly in place between the washers and the board.

5 Create a holder for the two batteries, like the one shown in figure 1, by hammering the 11 nails about a half inch (1.2 cm) into the board to surround the batteries. Don't forget the nail between the two batteries; this will allow you to use either one or two batteries to power the light bulbs. Point the negative ends of the batteries (the flat ends) toward the left.

6 Cut three individual bulbs from the string. Cut the wire halfway between adjacent bulbs so that each bulb ends up with two wire leads that are equal in length.

7 Use the wire stripper to strip about a half inch (1.2 cm) of insulation from the ends of the wires coming from each of the light bulbs.

8 Strip the insulation from the ends of each of the three additional pieces of wire.

9 Use the needle-nose pliers to attach alligator clips firmly to both ends of two of the pieces of wire (leave the other piece alone). A good way to attach the clips is to poke the stripped part of the wire down through the hole near the end of the clip so that the insulated part lies in the curved part of the clip between the two tabs. Then bring the stripped part up around the clip and lay it on top of the insulated part. Finally, bend the two tabs down on top of both the bare and the insulated parts to hold them tightly in place. This should ensure both a good electrical connection and a good physical connection (see figure 3).

Figure 3

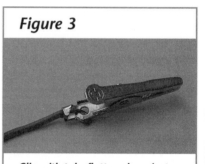

Clip with tabs flattened against both the bare and insulated wire

To Do and Notice

Making Connections

Refer to figure 4 (on page 11) for clip numbers, then connect a single bulb to clips 6 and 7. Use the alligator-clip leads you made to connect the nails at the ends of the batteries to clips 2 and 3. Connect clips 4 and 8 with a straightened-out paper clip, and use the plain piece of wire to connect clips 1 and 5. What happens to the bulb when you make the final connection? Your completed circuit should look like the opening photo. (**NOTE:** If the bulb doesn't light, see Helpful Hints on page 11.)

Connecting Bulbs in Series

Replace the wire between clips 1 and 5 with a bulb, and replace the paper clip between clips 4 and 8 with the third bulb. All three bulbs should light and glow with approximately the same brightness. (If one of the bulbs is very much brighter or dimmer than the other two, remove it and replace it with another bulb.) How does the brightness of these bulbs compare to the brightness of the single bulb in the initial setup?

Remove one of the bulbs from the circuit, and don't replace it with a wire or paper clip. What happens to the other two bulbs? Try replacing the bulb you removed, and removing another one instead. Does it matter which one of the bulbs you remove?

Changing the Voltage

Set up the circuit in its original form, with just one bulb. Move one of the alligator clips from a nail at the end of the battery holder to the nail between the batteries. What happens?

Connecting Bulbs in Parallel

Remove all bulbs, wires, and paper clips. Connect one of the alligator-clip leads from the end of the battery holder to clip 1, and the other lead from the other end of the battery holder to clip 6. Connect bulbs between clips 1 and 5 and between clips 2 and 6. What happens? Remove one of the bulbs. What happens now?

Replace the bulb you removed, and remove the other one. What happens? Does it make any difference if you remove one bulb or the other? How does the brightness of the single bulb compare with the brightness of each of the two bulbs? How does the behavior of the two bulbs differ from that of the three bulbs that were connected differently?

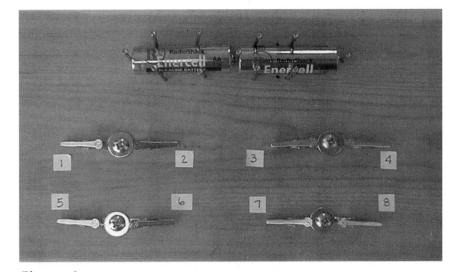

Figure 4 Board with clips numbered

What's Going On?

Connecting Bulbs in Series

In the initial setup with one bulb, electrons come out of the negative end of the left battery to begin their journey through the circuit. They travel through clips 2 and 1, the wire, clips 5 and 6, the bulb, clips 7 and 8, the paper clip, clips 4 and 3, and back into the positive end of the right battery. This path is called a complete electrical circuit. The flow of electrons in a complete circuit is often referred to as electrical current, although this is not technically correct—see the Current Versus Electron Flow discussion in the Did You Know? section.

In the batteries, the electrons gain energy. Batteries are rated in volts, and volts are a measure of the energy that electrons have when they leave the battery. We'll assume that the only place the electrons lose energy is in the bulb, where their energy is transformed to the light and heat given off by the bulb. The process of energy transformation resists the electron flow through the bulb, and the bulb is said to have electrical resistance. (If the battery voltage stays the same, changing the resistance does not change the energy of individual electrons, it just changes the number of electrons that flow. That is, it changes

the current.) It's very important to realize that while an electron loses energy on its trip around the circuit, the electron itself doesn't leak out of the circuit or disappear. For every electron leaving the battery to enter the circuit, another enters the battery from the circuit.

When two or more bulbs are connected so that electrons have no choice but to pass through all the bulbs to get back to the battery, the bulbs are said to be connected in series. When three bulbs are connected in series, each bulb glows less brightly than one bulb alone. This is because three bulbs provide more total resistance than one bulb, decreasing the current and resulting in a smaller amount of energy being transformed to light and heat. Additionally, each bulb only gets one-third of this reduced amount of energy.

When you remove one bulb in a series circuit, the others go out. It doesn't matter which one you remove; as long as you create a gap in the circuit, the electrons can no longer flow.

Changing the Voltage

When you move one of the battery clips to the nail between the batteries, the circuit is powered by one battery instead of two. This decreases the voltage by half, so the bulb gets dimmer.

Connecting Bulbs in Parallel

When you connect bulbs between clips 1 and 5 and between clips 2 and 6, the electrons arriving at clip 1 now have two alternative paths to get to clip 6. If the two light bulbs are identical, half the electrons will go through one bulb and half through the other. The electrons will then recombine into one flow at clip 6, and return to the battery. Bulbs connected in this fashion are said to be connected in parallel.

When two or more identical bulbs are connected in parallel, all the bulbs will be as bright as a single bulb. This is because each additional bulb provides another path for electron flow, allowing the battery to send more electrons through the circuit and giving each bulb the same amount of current as a single bulb. The alternative paths actually reduce the total resistance in the circuit. In fact, the resistance of two identical bulbs in parallel is half the resistance of a single bulb, allowing twice as much current to flow in the circuit. When one of the bulbs in the parallel circuit is removed, current still flows through the other bulb and it stays lit.

Summary

After you completed the experiments, you probably noticed the following differences between series and parallel circuits: Removing any one of the bulbs in a series circuit makes all the bulbs go out, but removing one or more of the bulbs in a parallel circuit has no effect on the remaining bulbs. Also, the more bulbs you place in series, the dimmer each one gets, whereas adding more bulbs in parallel has no effect on the brightness of any of the bulbs. (This last statement is true as long as the battery you are using can supply the electron flow required to light all the bulbs at their normal brightness. Although AA, C, and D batteries all have the same voltage [1.5 volts]—meaning that an electron leaving any one of them has the same energy—a AA battery is not capable of supplying as large a flow of these electrons as a D battery.)

So What?

Do you think the circuits in your house are wired in series or in parallel? If you have a toaster and a radio plugged into the same outlet, does the radio go off when the toast is done? The fact that you can turn electrical items in your house on and off individually tells you that your circuits are wired in parallel. If this were not the case, when a light bulb burned out, all other devices on the same circuit would go off.

It's obviously practical to have household appliances wired in parallel. But remember that with each additional device wired into a parallel circuit, the overall resistance of the circuit is decreased, allowing more current to flow. A sufficiently large amount of current can generate enough heat to melt the insulation on electrical wires and start a fire. To safeguard against this potential danger, we use circuit breakers or fuses. Circuit breakers are designed so that when the current exceeds a certain level, a switch opens to stop the flow. Fuses accomplish the same interruption of current flow by "blowing" (actually melting), which breaks the circuit.

Did You Know?

Current Versus Electron Flow

In some high school physics textbooks, and in most college physics and engineering texts, the direction of flow of electric current is defined as the direction positive charge would flow, which would be through the circuit from the positive terminal of the battery to the negative terminal. In reality, however, positive charges do not move in wires. It's the negatively charged electrons that move, and they flow through a circuit from the negative terminal of the battery to the positive terminal. The definition of current as movement of positive charge has historical roots, and continues in use; current defined in this way is often called conventional current. In some situations (e.g., move-ment of charged particles in semiconductors, solutions, and gases) positive charges do actually move and contribute to electric current. For beginning work with simple circuits, it's easier to deal with electron flow; in this snack we've used current and electron flow interchangeably.

Going Further

A Series of Connections

Think of a way to connect a third bulb in the circuit so that all three bulbs will retain their maximum brightness, and any one or two of them can be removed without the third one going out.

Experiment further with your circuit board. How many bulbs can you connect in series? How many in parallel? Try putting a bulb in series with two bulbs that are in parallel; this is called a series-parallel circuit.

Try some arrangements of your own design. Notice the paths that the electrons can take in the circuits that you set up, and see if you can understand the behavior of the bulbs in these circuits.

Measure It

Obtain a voltmeter and ammeter, and measure the voltage and current in the various circuits you set up. (You can buy an inexpensive multimeter, part number 990177, from Kelvin, 800-535-8469, www.kelvin.com.)

Circuit Equations

Look up Ohm's Law and Kirchoff's Rules for voltage and current in a physics text or other reference source, and see if you can apply them to your circuits.

Credits & References

This snack is based on the Exploratorium exhibit of the same name.

Mims, Forrest M. III. *Getting Started in Electronics*. RadioShack, 2000. (RadioShack #62-5004)

Diamagnetic Repulsion

Push around a grape without touching it.

What will happen if you bring an iron-containing piece of metal, such as a nail or a paper clip, near the pole of a magnet? It will be attracted, of course. Certain materials, however, behave very differently in magnetic fields: They are actually repelled by magnetic poles. These materials, which include the common substance water, are called diamagnetic. In this snack, you use the strong magnetic pole of a rare-earth magnet to repel a water-containing— and therefore diamagnetic—object.

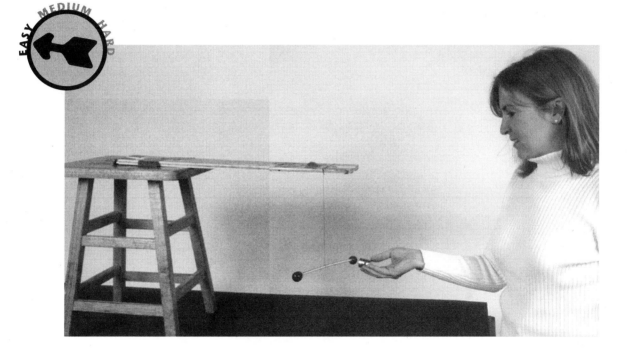

Materials

- thread, approximately 2 ft (60 cm)
- drinking straw (straight straw, without a flexible bend)
- 2 grapes that are about equal in size

- neodymium magnet (Only a neodymium magnet will be strong enough to work with this activity.) You can purchase a neodymium magnet inexpensively from Educational Innovations, 888-912-7474,

www.teachersource.com (#M-100, 0.75 in × 1.3 in × 0.47 in thick) or The Magnet Source, www.magnetsource.com. You may also be able to obtain one from magnet or scientific supply companies.

1 Tie one end of the thread in a loop around the straw with a slipknot (see figure 1) somewhere near the middle of the straw. Pull the loop as tight as possible without crushing the straw.

2 Push a grape onto each end of the straw.

3 Hold the thread so that the straw is suspended by about 12 inches (30 cm) of string. Tape or tie the thread to a

Figure 1

Thread and straw

Slipknot

Tie a slipknot around the straw.

suitable object so that the straw is suspended by at least a foot of thread (two feet is better if it can be arranged), and can rotate freely without hitting anything. The introductory photograph shows one possibility for improvising a way to hang the straw. Other ways would be to tape the thread to the edge of a table, or build a stand from PVC or wood.

4 Slide the loop of string along the straw until the straw is balanced as well as possible.

To Do and Notice

Bring one pole of the magnet near one of the grapes, as shown in figure 2. It's important that the magnet is *very near* the grape, but that it *doesn't touch* the grape.

The grape will be repelled by the magnet and will begin to move slowly away from it.

Remove the magnet and let the grape stop its motion. Turn the magnet over and bring the other pole near the grape. Does the other pole of the magnet affect the grape in the same way?

What's Going On?

The grapes contain the diamagnetic substance, water. Water, like all diamagnetic materials, is repelled by both poles of a magnet because of the structure and behavior of the electrons in its molecules.

Magnetic attraction and repulsion come from the movement of electrons in matter. Electrons move in two ways—they spin and they move around the atom's nucleus. Because electrons carry charge, both kinds of movement create magnetic fields.

The magnetism you are probably used to seeing is *ferromagnetism*. This is the force that holds magnets to refrigerators. In ferromagnetic materials (iron, cobalt, and nickel), the spin of electrons in one atom lines up with the spin of electrons in neighboring atoms. These aligned electrons are strongly attracted to magnetic poles.

In this snack, you are experimenting with *diamagnetism,* which is so much weaker than ferromagnetism that most people never notice it. Diamagnetism is actually a property of all matter. As explained by Lenz's law

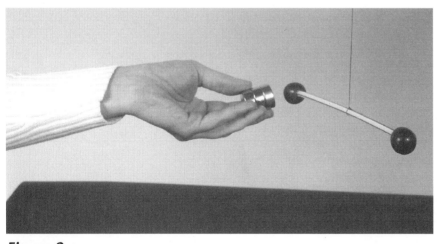

Figure 2 Hold the magnet as close as you can to the grape without actually touching it.

(see box below), the application of a strong external magnetic field generates orbital electric currents in atoms and molecules. The magnetic fields associated with these orbital currents are oriented such that they are repelled by the approaching magnet. This repulsion is relatively weak, and in most materials it is overcome by much stronger forces of ferromagnetism and *paramagnetism.*

Paramagnetism is much weaker than ferromagnetism and stronger than diamagnetism. Paramagnetism occurs in substances made up of atoms or molecules with single, unpaired electrons. When subjected to a strong magnetic field, these electrons don't align with each other as in ferromagnetism. Instead, they orient themselves in the magnetic field so that they are slightly attracted to the magnetic poles. This attraction, though weak, is still stronger than the underlying diamagnetic force of repulsion. Many substances are paramagnetic, including hydrogen, lithium, and liquid oxygen.

Some substances don't display ferromagnetism and paramagnetism. In these substances, you can see the effects of diamagnetic repulsion. In these substances, all the electrons in the atoms or molecules are paired with electrons of opposite spin. These diamagnetic materials include water, helium, and bismuth.

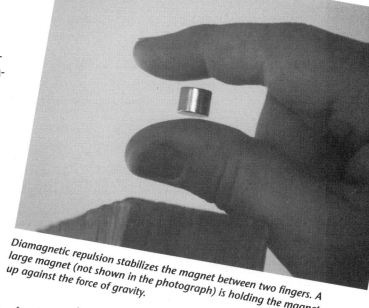

Diamagnetic repulsion stabilizes the magnet between two fingers. A large magnet (not shown in the photograph) is holding the magnet up against the force of gravity.

This frog is floating, suspended by diamagnetic forces from a huge magnet. The magnet is not shown in the photograph.

So What?

Diamagnetism can be used to levitate objects, such as fruits, nuts, and balls of water. Scientists recently levitated a frog—with no harm to the animal. Diamagnetic levitation mimics the effects of weightlessness—however, only the strongest, largest magnets are capable of levitating an object, such as a frog, against the gravitational attraction of the earth.

Going Further

Fruitful Possibilities

Try fruits other than grapes. Fruits with a high water content, such as watermelons, work well. Cut large fruits into chunks the size of grapes. Prunes will also work and have the advantage of not spoiling.

Paramagnetic or Diamagnetic?

You can experiment with materials such as graphite, aluminum, and wood. If these materials are attracted, they are exhibiting paramagnetism. If they are repelled, they are exhibiting diamagnetism.

Credits

We first saw this phenomenon demonstrated by the Galileo Circle, a group of Japanese science teachers.

Heinrich F. E. Lenz

1804–1865

A professor of physics at the University of St. Petersburg in Russia, Lenz studied electrical induction in the 1830s. He described the phenomenon of self-induction in the principle known as Lenz's law (see What's Going On?). Simply put, Lenz's law states that an electric current created by a changing magnetic field will create its own magnetic field that opposes the charge in the magnetic field that created it.

Film Can Racer

A new version of an old toy.

The Film Can Racer is a modern adaptation of an old toy, the spool racer. Though simple to make and use, it illustrates a number of physics principles—and it's fun!

Materials

- electric drill
- drill bit, $\frac{1}{4}$ in or slightly smaller
- film canister with lid

- utility knife or X-Acto knife
- 3 rubber bands, a little longer than the length of the film canister (e.g., size 18)

- small hex nut (e.g., $\frac{1}{4}$ in or 10-24)
- drinking straw (or pencil, wooden dowel, or bamboo skewer)

1 Drill a hole in the center of the bottom of the film canister. The hole must be large enough for a rubber band to fit through without rubbing, but small enough to keep the hex nut from going through.

2 Cut a flap in the lid as shown in figure 1. Film canister lids have various shapes, depending on the brand of film, and three different lids with flaps cut are shown here as examples.

Figure 1

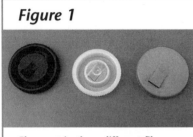

Flaps cut in three different film canister lids

3 Push one end of a rubber band through the hole in the bottom of the canister so that most of the rubber band is inside. Thread the end of the rubber band that's outside the canister through the hex nut.

Figure 2

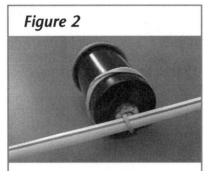

The rubber band goes through the hole in the bottom of the film canister, through the nut, and around the straw.

Push the straw through the loop of rubber band that sticks out through the hex nut as shown in figure 2.

4 Grasp the end of the rubber band that's in the canister and push it through the hole in the lid created by the flap. Slip the loop of the rubber band around the flap and move it toward the hinge of the flap, and then put the lid on the canister (see figures 3 and 4).

5 Wrap one or two rubber bands around the canister near the bottom to match the rim created by the lid on the other end.

Figure 3

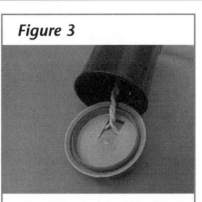

The other end of the rubber band is wrapped around the flap.

Figure 4

Completed Film Can Racer

To Do and Notice

Turn the straw to wind up the rubber band, but don't wind it too tightly. Put the racer on the floor, release it—and watch it go!

If you wind the rubber band tighter, does the racer go farther? (You might try counting the number of turns you wind the rubber band, so that you can make comparisons.) What happens when you wind the rubber band extremely tight?

Run the racer on different surfaces (for example, wood, carpet, concrete). Does the surface affect the racer's performance?

Try using different rubber bands (e.g., thinner, thicker, shorter, longer). How does the type of rubber band affect the performance?

Wind the straw, pick up the racer, and release the straw while holding the canister. Note that the straw will spin like a propeller when released. What happens when you hold the straw and release the canister?

Why do you need the rubber bands that form the rim? What happens if you take them off? Try it and see.

The Film Can Racer is similar to a toy that children used to make from wooden spools.

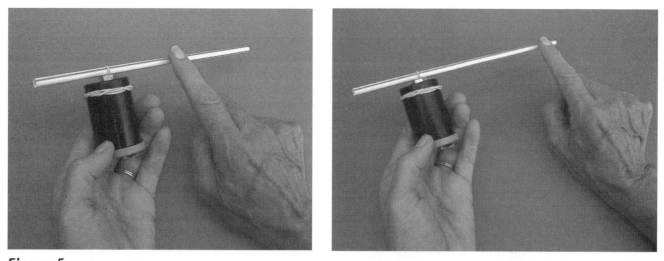

Figure 5 *It takes more force to push the straw when your finger is close to the film canister than when it is farther away.*

What's Going On?

When you turn the straw to wind up the rubber band, you exert a force on the end of the straw and move it through a distance—which means that you are doing mechanical work. This mechanical work is stored in the rubber band as potential energy. When you let the racer go, the potential energy is transformed into kinetic energy.

When you rotate the straw, you are exerting a *torque,* or twist. The farther out on the straw you push, the less force you need, but the greater the distance through which you must move the force (see figure 5). No matter where you push, the work done (the product of force times distance) is the same—but you can choose whether you want to use a large force through a small distance or a small force through a large distance.

When you release the racer, the straw exerts a force on the floor. The floor exerts a reaction force back on the straw, so the straw itself can't rotate. This forces the canister to rotate instead (provided the rubber band is wound tightly enough to overcome friction), and the racer moves. (Remember that when you wound up

the straw, held it, and then released the canister, the canister rotated.)

The wound-up rubber band produces a torque on the lid. This torque originally consists of a relatively large force located a short distance from the center of the canister. The torque remains constant and reappears at the edge of the canister as a smaller force located a larger distance away from the center. This force acts perpendicular to the radius of the canister—that is, tangent to the edge—where the canister is in contact with the floor. The reaction to this force is the tangential force that the ground in turn exerts back on the racer, and it is this force that propels the racer forward.

When you take off the rubber bands that form the rim at the bottom end of the canister, that end of the can has a smaller circumference than the other. Every time the canister makes one revolution, the lid

end goes farther than the bottom end, causing the racer to travel in a circular path rather than in a straight line.

So What?

Newton's third law of motion states that "for every action there is an equal and opposite reaction." When a car goes forward, for example, each tire exerts a force backward on the road at the point of contact. (If there were loose gravel on the road, and the car "spun out," pieces of gravel would go flying backward.) If we call the force that a tire exerts on the road the "action," then the "reaction" to it is the road pushing on the tire with an equal force in the opposite direction. This is the force that makes the car go forward as shown in figure 6.

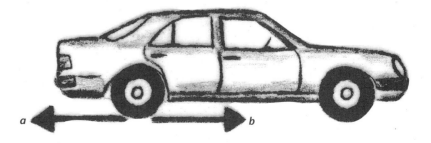

Figure 6 *(a) Action force: tire pushes on road; (b) reaction force: road pushes on tire*

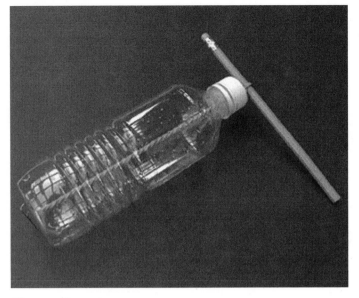

Figure 7 *Soda bottle version*

Did You Know?

Distance Versus Displacement

Distance (total length that the racer travels) and displacement (how far away from its starting point the racer ends up) are two distinct concepts in physics. A racer that travels completely around a large circle has traveled a long distance but has zero displacement! If you are trying to maximize the displacement of your racer, it's important that you make it travel in a straight line.

Going Further

Racing Racers

Try to optimize the performance of your racer for either distance or speed. If you have friends who want to build racers, you can set up a competition to see whose racer will travel the farthest or travel a given distance in the shortest time.

Bottles and Cans

You can also make racers out of plastic bottles (see figure 7) or coffee cans. Some helpful hints:

- *Plastic bottles* Drill a hole in the bottom; get a larger rubber band; use a paper clip to keep the end of the rubber band from slipping into the bottle; make a wire hook out of a coat hanger to reach inside the bottle to pull the rubber band through.

- *Coffee cans* Replace both metal ends with the plastic lids that come with coffee cans.

Gearing Up . . . and Down

If you make racers out of bottles or cans, try making racers with different diameters. Other things being equal, a larger diameter bottle exerts a smaller tangential force on the ground, but its larger circumference means that it travels farther in one revolution. This is like high gear. Conversely, a smaller bottle exerts a larger force, but one revolution doesn't get it very far. This is like low gear. Compare the speed and hill-climbing abilities of your differently "geared" racers.

References

Hewitt, Paul. *Conceptual Physics,* 8th ed. Menlo Park, Calif.: Addison Wesley, 1998. There are nice discussions of Newton's third law of motion on pages 70–75 and of torque on pages 121–122.

Isaac Newton

1642–1727

Sir Isaac Newton, a brilliant English physicist and mathematician, developed the three laws of motion that we still use today. A discussion of these laws can be found in any physics book.

In addition, Newton invented calculus, formulated the law of universal gravitation, and demonstrated that white light is composed of many colors. In 1687, Newton published his discoveries of the laws of motion and theories of gravitation in *Philosophiae Naturalis Principia Mathematica* (Mathematical Principles of Natural Philosophy), which is generally considered to be one of the greatest scientific books ever written.

Fractal Patterns

Make dendritic diversions and bodacious branches.

Lightning bolts, river deltas, tree branches, and coastlines are all examples of fractals—irregular patterns that repeat themselves at different scales. In this snack, you get a striking hands-on introduction to fractal patterns and how they are formed.

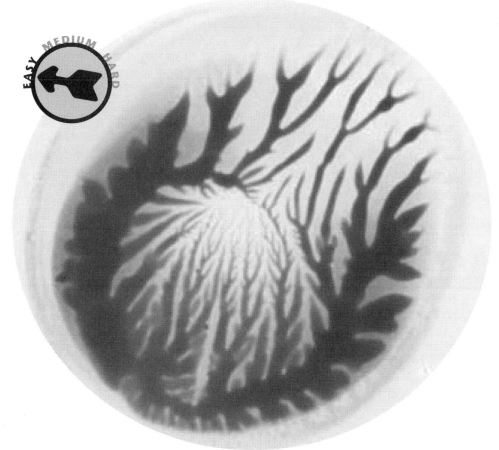

Materials

- paper clip, toothpick, bamboo skewer, or nail
- gloss enamel model paint (e.g., Testors Gloss Enamel), assorted colors. Liquitex water-based model paint works almost as well, is nontoxic, and cleans up with water, but it takes longer to dry.

You can also try acrylic paint, poster paint, frosting, frosting gels, margarine, latex paint, or other materials—each has its advantages and disadvantages.

- 2 same-size pieces of clear plastic, the smoother and more rigid the better, 1 in × 1 in (2.5 cm ×

2.5 cm) to as large as 4 in × 4 in (10 cm × 10 cm); examples: CD jewel cases, #6 plastic used for deli containers that can be cut with scissors, or clear acrylic plastic; the pieces of plastic don't have to be square.

No assembly needed	## Alternative Construction
	Cover your pieces of plastic with transparent tape so you can peel off the fractal pattern when the paint is dry. This allows you to reuse the plastic, as well as to tape the pattern to a piece of paper or use it in some other creative way. The tape has to be wide enough to allow the pattern to be created on a single piece; two-inch-wide transparent packaging tape works well.

To Do and Notice

1 Use a straightened-out paper clip, a toothpick, a bamboo skewer, or a nail to stir the paint. Then use the same implement to place a tiny drop of paint at the center of one of the plastic pieces, which are your plates.

2 Place the second plate on top of the paint, but don't line up the edges of the top plate exactly with the edges of the bottom plate. (You will be pulling the plates apart, and if the edges are lined up, this may be difficult; see figure 1, step a.)

3 Squeeze the two plates together firmly, so that the paint drop forms the thinnest possible layer between them. Notice that the paint spreads into a disk (see figure 1, step b).

4 Carefully pull the plates apart as shown in figure 1, step c. *Do not slide them apart.* It's very important that you pull the plates straight off one another. Watch air flow into the paint as you pull the plates apart, forming a fractal pattern.

5 Once the plates are separated, observe the patterns on each one. Notice that the patterns are mirror images of each other. What do you think the patterns look like? Do they remind you of anything? Let the paint patterns dry on the plates if you want to preserve them.

6 If you have access to an overhead projector, try projecting the patterns on a white wall or screen. You may not get any color unless you are using gels, but the outlines should still be impressive.

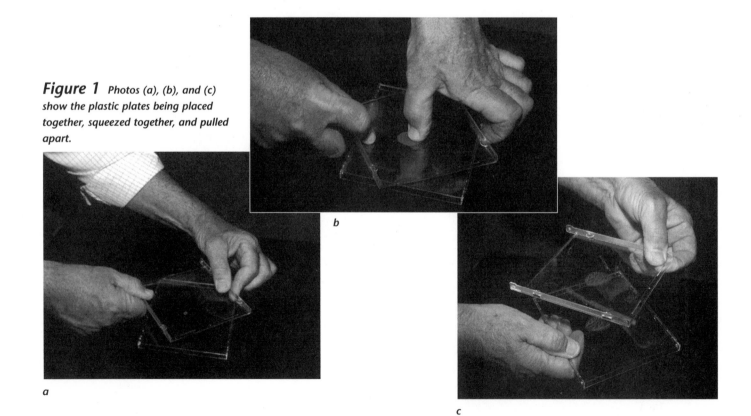

Figure 1 *Photos (a), (b), and (c) show the plastic plates being placed together, squeezed together, and pulled apart.*

b

a

c

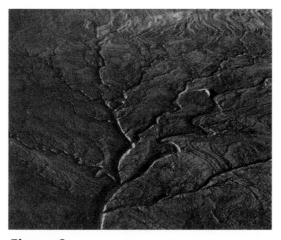

Figure 2 *This aerial photo of dry streambeds in the California desert shows an example of a dendritic pattern—a pattern with many branches or forks.*

What's Going On?

Your fractals are the result of a process called *viscous fingering:* As the paint is squeezed between the plates, the viscous paint spreads out evenly in all directions into the less viscous air layer, creating a stable, disk-shaped boundary.

When the plates are pulled apart, the less viscous air penetrates the more viscous paint, creating an unstable boundary. Small indentations of air grow and become fingers of air. Random indentations in these fingers grow as well. By the time the two plates are separated, the fingers of air have formed intricate branching structures in the paint.

The patterns created with this process often remind people of such things as tree branches or root systems, river deltas, and lightning bolts, all of which are outstanding examples of fractal patterns in nature. A few others are clouds, coastlines, jellyfish tendrils, coral reefs, and blood vessels in the lungs. Figure 2 shows another example of a natural fractal pattern, the branching of a river. This is typical of the pattern formed when fluid flows from tributaries into a central stream or flows out from a main course into smaller branches.

All the fractal patterns formed in nature—including the ones you just made—are generated by random processes. As the patterns repeat themselves at different scales, each section of the whole is similar to large and smaller scale structures, but never an exact copy. If you break a floret of cauliflower or broccoli off the larger head, for example, you can see that it's like a miniature version of the larger head, but it's not an exact replica. These repeating but nonidentical patterns are called *self-similar.*

Even though nature cannot generate a perfect fractal pattern, mathematicians can. The fractals they create are called *perfect* or *mathematical* fractals. If you look at a small section of a mathematical fractal, the section will be identical to the whole object. Furthermore, if you were to select an even smaller piece from the first section and magnify it, this piece would also duplicate the whole. In fact, in this kind of fractal, you can't tell the difference between the whole object and a magnification of any section you select—no matter how small! Figure 3 shows a classic mathematical fractal, the Sierpinski Triangle, or Sierpinski Gasket.

Figure 3 *Sierpinski Triangle*

Benoit B. Mandelbrot

1924–

Benoit Mandelbrot, a Polish-born French mathematician who now lives in the United States, is credited with the development of fractal geometry. His various studies have included stock-market fluctuations, the turbulent motion of fluids, and the distribution of galaxies.

So What?

Fractals help us to understand many different areas of science, including crystal growth, earthquake processes, meteorology, and polymer structure, to name just a few. Fractals are particularly significant in the field of *chaos theory,* which seeks to explain apparently random behavior that occurs within a system.

Fractals have also had a profound impact on computer graphics, with stunning fractal images forming a significant new interface between science and art. Moviemakers have used fractal imaging techniques to create special effects in movies such as the *Star Wars* trilogy and *Star Trek: The Wrath of Khan.*

Did You Know?

A Broken Word

The word *fractal* was first used in 1975 by Benoit Mandelbrot. The word derives from the Latin *frangere* (to break) and *fractus* (broken, uneven).

Going Further

Fractal Inspiration

Make a fractal pattern on clear tape (see Alternative Construction). Stick the resulting two pieces of tape to a piece of paper, and create art or poetry inspired by the fractal images.

Assembly Instructions

The Sierpinski Triangle (see figure 3) and the Cantor Set, or Cantor Dust, are two classic fractals whose algorithms (the set of instructions for their construction) can be easily understood. Look these up on the Web to get an idea of how a mathematical fractal is generated.

Credits & References

This snack was developed by Linda Shore and the educational staff at the Boston University Center for Polymer Studies.

Barnsley, Michael. *Fractals Everywhere.* San Diego: Academic Press, 1988.

Mandelbrot, Benoit. *The Fractal Geometry of Nature.* San Francisco: W. H. Freeman, 1982.

Peitgen, Heinz-Otto, and Peter H. Richter. *The Beauty of Fractals: Images of Complex Dynamical Systems.* Berlin: Springer-Verlag, 1986.

Hoop Nightmares

Retrain your brain.

When you first try shooting a basket or throwing a ball at a target, you'll probably come pretty close, even if you might not consider yourself quite ready for the pros. But when you put on this special set of prism goggles and try to make the same shots, things get very interesting.

Materials

- plastic safety goggles with a flat face plate (available at hardware and home improvement stores or from scientific supply companies)
- dark-colored poster board, approximately 4 in × 8 in (10 cm × 20 cm), large enough to cover the face plate of your goggles
- pencil
- scissors
- masking tape

- plastic or glass prism, about 2 in (5 or 6 cm) long, with faces about 1 in (2 or 3 cm) wide; angles should be 45/45/90 or 30/60/90 (You can obtain prisms from Edmund Scientific Co., 800-728-6999, www.edsci.com, e.g., #30318-00, science museum stores, science supply companies, and some novelty or magic stores; long plastic prisms can be cut to shorter lengths with a band saw or hacksaw.)

- a hoop or a target to throw the ball at (A Nerf hoop is ideal, but you can improvise a hoop from a coat hanger or use a cardboard box on a chair or table.)
- a ball that can be thrown indoors without breaking things (A Nerf ball is ideal.)

1 Lay the safety goggles face down on the poster board, trace the outline of the goggles on the poster board, and cut out the outline to create a poster-board mask.

2 Cut a rectangular opening in the mask just to one side of the middle of the goggles, as shown in figure 1. You will be taping the prism over this hole, so its size will depend on the size of the prism you have. A general guideline is to make the width of the opening half as wide as the widest face on the prism.

Figure 1

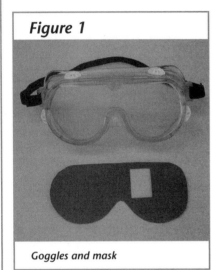

Goggles and mask

Figure 2

Goggles with mask and prism in place

3 Tape the prism in place over the hole on the poster-board mask. The edge of the prism nearest the middle of the mask should just overlap the edge of the hole, as shown in figure 2. Put the largest face of the prism over the hole. If you're using a 30/60/90 prism, the next largest face should be facing the center of the mask. If you're using a 45/45/90 prism, one of the equal-size faces should be facing the center of the mask (you can try other orientations later to see the difference, if you want to).

4 Tape the entire mask, with the prism attached, to the front of the goggles. When you put the goggles on, the prism will be in front of your left eye.

5 Cover the face of the prism that will be farthest from your nose with masking tape, as shown in figure 2, so that only the face of the prism slanting toward the middle of the goggles will receive light.

6 Set up a hoop or set up another target.

To Do and Notice

Stand about 9 or 10 feet (3 m) from the hoop or target. (The farther you are from the target, the more obvious the effect will be.) Put the goggles on, and make sure that you can see the hoop or target through the goggles. (You will only see it with one eye, because there is only one prism.) **NOTE:** Make sure that your throwing hand is positioned so that you cannot see it or the ball through the goggles.

Try to make a basket or hit the target with the ball. Notice how close you get. Have your partner retrieve the ball for you if possible—it's hard to do with the goggles on.

Keep trying until you make a basket or hit the target three times in a row. How many tries does it take you?

Take the goggles off and try again. Keep trying until you score three times in a row. How many tries does it take you?

What's Going On?

When you first put on the goggles, the ball doesn't go where your eye says it should. Because of the way it *refracts,* or bends, light, the prism makes objects in front of you appear to be to your right (see figure 3). Your brain, however, soon adapts to the distortion and adjusts your visual perception. You begin to aim farther left and get closer to hitting the target or hoop.

When you remove the goggles, your visual system remembers the prism distortion, and it functions as if the goggles were still in place. It may take a few trials for your visual system to "unlearn" the adjustments it made and return to normal.

Your experience with the goggles shows that your brain and perceptual systems are dynamic: They continually respond and adapt to your experiences, whether or not you're thinking about them.

How does the prism distort your view of the world? Light travels from the hoop (or target) to your eye along the path shown by the arrows in figure 3. As it passes through the prism, it is bent twice—once when it enters the prism and again when it leaves. Your eye-brain system tries to follow this light back to its origin in order to locate the hoop, but it doesn't have the ability to recognize that the light was bent. It follows the light back along a straight line defined by the ray of light that enters your eye, and so the hoop appears to be somewhere on this line.

Figure 3 *The solid arrows show the path of light from the hoop to your eye. The dotted line shows where your eye-brain system sees the hoop.*

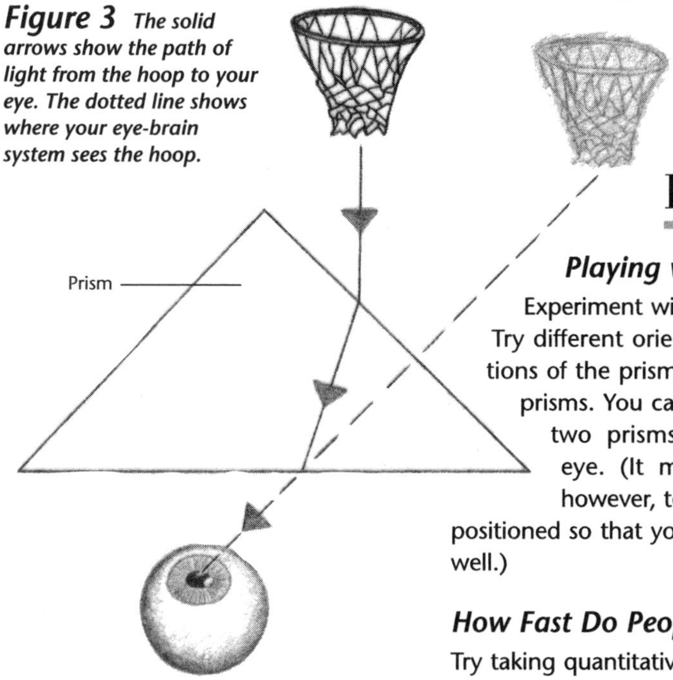

Prism

So What?

You forced your visual system to adapt when you put on the goggles. But it doesn't take an artificial and extreme action like looking through a prism to bring this ability to adapt into play.

On a daily basis your brain is challenged to relearn skills and change the way it processes information—all it takes is driving someone else's car, taking a new route to the grocery store, or putting your toothbrush in a new location. If we did not have the ability to adapt to changes in the world (or to changes in our perception of the world), life would be much harder.

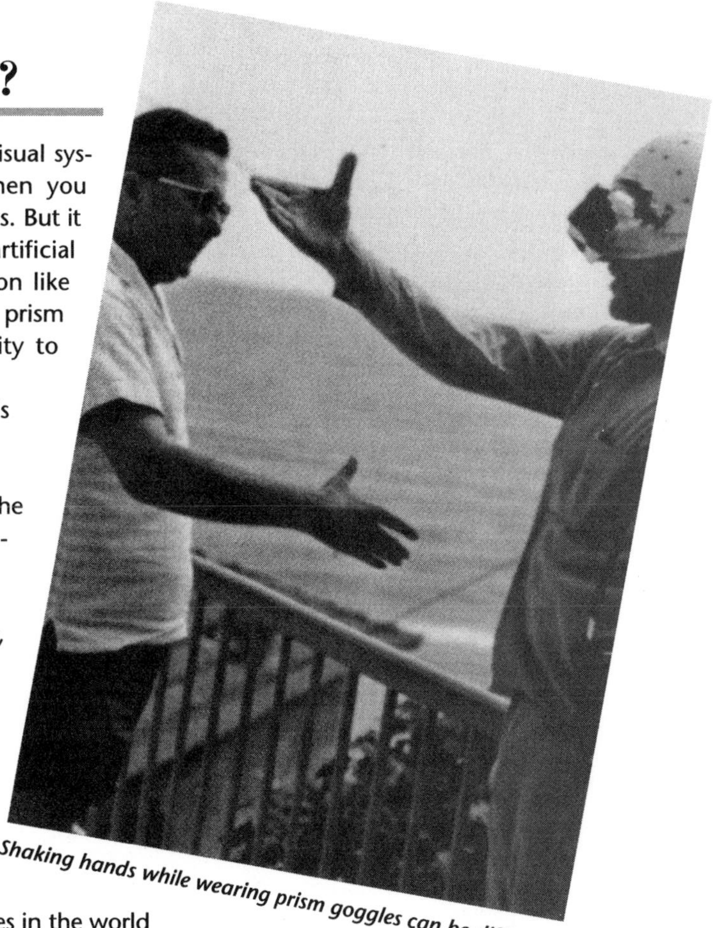

Shaking hands while wearing prism goggles can be difficult.

Going Further

Playing with Prisms

Experiment with your goggles. Try different orientations or locations of the prism or use different prisms. You can even try using two prisms, one for each eye. (It may be difficult, however, to get two prisms positioned so that you can see things well.)

How Fast Do People Adapt?

Try taking quantitative data for different people using the prism goggles to find the range of the learning and unlearning times.

Old Habits Die Hard

Is there something in your house that you use a lot and that has been in the same location for a long time? Change its location, and notice how long you reflexively keep trying the old location first. How long does it take for you to completely change?

Credits

This snack is based on the Exploratorium exhibit of the same name.

Hydraulic Arm

You'll feel the pressure to do some heavy lifting.

When you push the plunger on a syringe, water is forced into a second syringe, extending its plunger and lifting a mechanical arm. The process illustrates aspects of fluid pressure, force, mechanical work, and biomechanics.

Materials

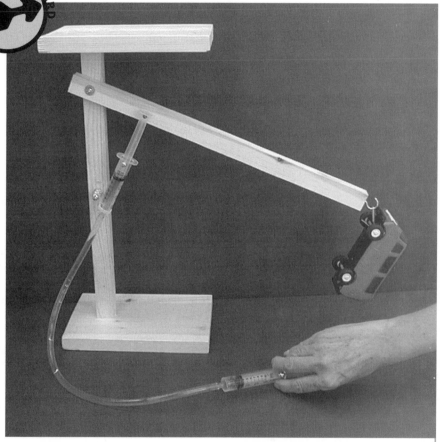

- electric drill
- drill bits, $\frac{3}{16}$ in, $\frac{1}{8}$ in, $\frac{1}{16}$ in
- 2 wooden bases, about 5 in × 8 in × $\frac{3}{4}$ in (12.5 cm × 20 cm × 2 cm) thick (ordinary 1 in × 6 in pine shelving works well)
- 1 wooden post, about 15 in × 1 in × $\frac{3}{4}$ in (38 cm × 2.5 cm × 2 cm) thick
- 1 wooden arm, about 16 in × 1 in × $\frac{3}{4}$ in (40 cm × 2.5 cm × 2 cm) thick
- 2 flat-head wood screws, #8 × 1$\frac{1}{2}$ in
- 2 machine screws, 10-24 × 2 in
- screwdrivers to fit screws
- 5 washers to fit the machine screws (e.g., SAE 10 flat washers)
- 2 wing nuts, 10-24
- cup hook, $\frac{7}{8}$ in or 1 in
- 2 plastic oral syringes, 10 mL (Oral syringes are used by pharmacists and veterinarians to accompany prescriptions and may sometimes be obtained for no charge or purchased inexpensively. You can also obtain syringes, tubing, connectors, etc., for hydraulic and pneumatic projects from Kelvin, 800-535-8469, www.kelvin.com, or from The Science Source, 800-299-5469, www.thesciencesource.com; see Alternative Construction for use of 5-mL syringes if you can't get the 10-mL size.)
- 1 oral syringe, plastic, 1 mL

- 2 sheet-metal screws, short (e.g., #6 × $\frac{3}{8}$ in)
- 1 nail, bright box, 1$\frac{1}{2}$ in
- 1 cable tie with mounting head, 7.5 in (usually available at home improvement stores and some hardware stores; see Alternative Construction using an eyebolt if you have trouble obtaining the cable tie with mounting head)
- pliers
- scissors
- 4 hex nuts, 10-24

- 2 ft (60 cm) of clear plastic tubing to fit the syringes you use (available at hardware and aquarium supply stores; the tubing that fit the syringes we used had an outside diameter of $\frac{5}{16}$ in, an inside diameter of $\frac{3}{16}$ in, and a wall thickness of $\frac{1}{16}$ in)
- water source
- assorted small objects that can be hooked or tied onto the cup hook of the arm (e.g., toys, set of keys)

Figure 1

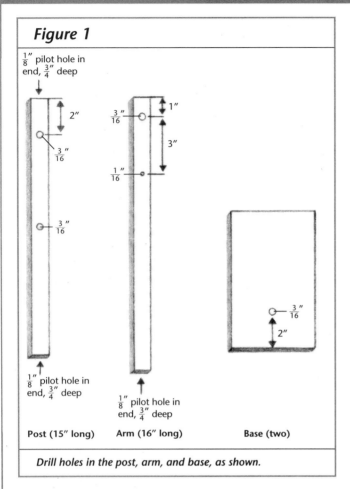

$\frac{1}{8}''$ pilot hole in end, $\frac{3}{4}''$ deep

2"

$\frac{3}{16}''$

$\frac{3}{16}''$

$\frac{3}{16}''$

$\frac{1}{8}''$ pilot hole in end, $\frac{3}{4}''$ deep

Post (15" long)

1"

$\frac{3}{16}''$

3"

$\frac{1}{16}''$

$\frac{1}{8}''$ pilot hole in end, $\frac{3}{4}''$ deep

Arm (16" long)

$\frac{3}{16}''$

2"

Base (two)

Drill holes in the post, arm, and base, as shown.

Figure 2

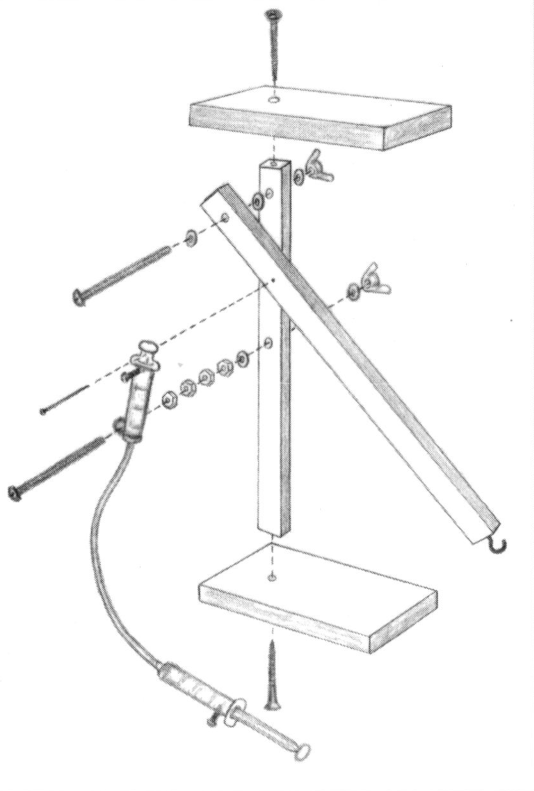

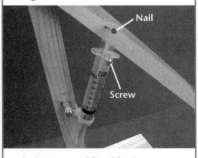

Exploded view of the whole assembly

1 Drill holes in the four pieces of wood as shown in figure 1. (There are two bases to allow you to use the hydraulic arm in the "upside down" position.)

2 Figure 2 shows an exploded drawing of the assembly of the Hydraulic Arm. Refer to this drawing as necessary as you proceed through the assembly steps that follow. (See figures 3 and 4 for details regarding the cable tie and stop screw.)

3 Assemble the two bases and the post, using the two wood screws. Be sure the heads of the wood screws do not stick out from the bases.

4 Attach the arm to the post, using a machine screw, three washers, and a wing nut.

5 Screw the cup hook into the end of the arm.

6 Drill a small hole (smaller than the sheet metal screws) in the body of each of the two 10-mL syringes, just below the top flange. Screw a small sheet-metal screw into each of these holes, and rotate the syringe plungers so that the screws can go in as far as possible without hitting the plungers (see figure 3). The screws should now act as stops to keep the plungers from being pulled completely out of the syringes by accident.

Figure 3

Nail

Screw

Syringe assembly with pivots

7 Select one of the syringes, and drill a $\frac{1}{8}$-inch hole through the shaft of its plunger approximately $\frac{1}{4}$ inch (0.5 cm) from the end of the plunger. A nail will later be placed through this hole (see the "Nail" label on figure 3). This syringe is the "fixed syringe."

8 Place the cable tie on this syringe as shown in figure 4. Pull it as tight as possible so the syringe does not slip easily. (If necessary, use pliers to pull the cable tie tighter after you have tightened it initially by hand.) Cut off all but about $\frac{1}{4}$ inch (0.5 cm) of the excess tie as shown in figure 5.

Figure 4

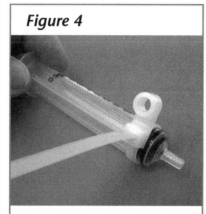

Cable tie on syringe

9 Attach the syringe to the post by placing a machine screw through the hole in the cable tie. Then thread onto the machine screw four hex nuts and a washer before attaching the syringe to the post (see figure 5) and adding another washer and a wing nut (see figure 2). Adjust the hex nuts to allow enough space between the last hex nut and the head of the machine screw so that the mounting head of the cable tie will allow the syringe to pivot freely but not slide sideways excessively. When this adjustment has been made, make sure that the four hex nuts are tight against each other, and tighten the wing nut.

10 Put the nail through the hole in the syringe plunger, and force it into the hole in the arm until the head of the nail is almost up against the plunger (see figure 3), but not so far that the point of the nail protrudes significantly from the other side of the arm. If necessary, use pliers to push the nail into the hole.

11 Attach the plastic tubing firmly to the tip of the second 10-mL syringe. Fill the syringe and tube completely with water. Eliminate air bubbles by flicking the syringe or tube with your finger to get them to rise, and then topping off with more water as necessary. If you have difficulty, you might try filling the tube separately and then attaching it to the full syringe.

12 Push the plunger all the way into the fixed syringe (the one attached to the post and arm) so that there is no air in the syringe. Attach the open end of the water-filled tube firmly to the tip of this syringe. Be sure there are no large air bubbles anywhere in the system.

Figure 5

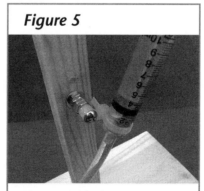

Pivot assembly close-up

Alternative Construction

- An eyebolt (plus two washers, and two hex nuts or two Teflon-insert stop nuts) can be used as an alternative to the cable tie for holding the syringe (see figure 6). Masking tape can be used to build up the syringe diameter to fit the eyebolt hole. The eyebolt shown here is $\frac{5}{16}$ inch by $3\frac{1}{4}$ inches, but hole sizes on eyebolts from different manufacturers are not uniform; be sure the hole on the eyebolt is large enough for your syringe to fit into.

Figure 6

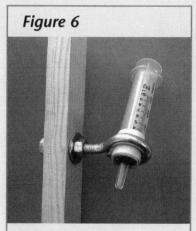

Eyebolt holding syringe

- If necessary, you can substitute 5-mL plastic oral syringes for the 10-mL ones. These are commonly sold in drug stores. They are slightly smaller, and the plungers do not extend quite so far. Depending on the size of the syringe you use, you may need to modify the locations of the holes used to fix the syringe to the arm and post.

To Do and Notice

Push on the plunger of the movable syringe. What happens? Pull on the plunger. What happens now?

Use the arm to lift a small object. (If the arm tips, either put a book or other heavy weight on the base to steady it, or find a lighter object to lift.) How does the force of your push on the plunger compare to the force that the other plunger exerts on the arm? How do the distances that the two plungers move compare with each other?

Notice carefully how hard you have to push on the plunger to lift a particular object, and notice how far the arm can move the object.

Remove the object, then push water into the fixed syringe so that the arm is elevated as much as possible. Support the elevated arm so that it can't fall. (You could have a friend hold the arm, or you could support it with a stack of books.) Then raise the movable syringe until the end of the tube attached to it is well above the fixed syringe. Keeping the end of the tube raised (to prevent water from coming out when the syringe is removed), remove the 10-mL syringe and replace it with the 1-mL syringe.

Remove the support from the arm. Pull the plunger on the 1-mL syringe until the syringe is full. Replace the object on the hook, and then push the plunger on the 1-mL syringe to lift the object.

Notice the difference in how hard you have to push on the plunger to lift the object this time, and notice how far the object is lifted.

Put the 10-mL syringe back in place at the end of the tube, using the same technique you used to replace it with the 1-mL syringe. Turn the whole device upside down and use the syringe to raise and lower the arm. Compared to the right-side up position, what's different about the process of elevating the arm?

Blaise Pascal

1623–1662

The pascal unit of pressure is named for Blaise Pascal, a French mathematician, scientist, and philosopher. His chief contribution to physics was in the field of *hydraulics,* which applies the properties of water and other liquids, such as how they transmit pressure, to engineering problems. In about 1650, Pascal authored a treatise, *On the Equilibrium of Liquids,* in which he states that an external pressure applied to a liquid in a container is transmitted equally throughout the container. This idea, which has come to be known as Pascal's principle, is the underlying physics of this snack.

What's Going On?

When you push on the plunger of the movable syringe, the arm rises; when you pull on the plunger, the arm descends.

Pushing on the plunger applies pressure on the water in the movable syringe. Since the water is confined and incompressible, Pascal's principle comes into play, telling us that the pressure is transmitted undiminished to all parts of the water and to the walls of its container. Since the plunger of the fixed syringe at the other end of the tube forms part of the container for the water and is the only part of the container that can expand, the pressure causes the plunger in the fixed syringe to move.

Pascal's principle and a little mathematics can be used to show that—if the syringes are identical—the force you apply to one plunger is transmitted in full to the other plunger (see Box o' Math). Additionally, as you can observe, each plunger moves the same distance.

With the 1-mL syringe, the force that you push with is noticeably less than that with the 10-mL syringe, but the arm is not lifted nearly as far. In accordance with Pascal's principle, the pressure on the plunger of the 10-mL syringe is the same as the pressure on the plunger of the 1-mL syringe. But since the area of the 10-mL plunger is far larger than the area of the 1-mL plunger, the force exerted on the 10-mL plunger is far larger than the force you push with (remember, $F = pA$).

Box o' Math

Pressure is defined as force per unit area

$$p = \frac{F}{A}$$

If you divide the force you push with by the area of the plunger that is in contact with the water, you can find the pressure exerted on the water. You can mathematically rearrange the equation above to become $F = pA$. This tells you that if you multiply pressure (expressed in pounds per square inch) by area (expressed in square inches) the square inches cancel out, and you are left with force, expressed in pounds.

$$\text{pressure} \times \text{area} = \frac{\text{pounds}}{\text{square inch}} \times \text{square inch} = \text{pounds}$$

Because the pressure on both plungers is the same, and the areas of both plungers are identical, then the force on both plungers is the same. (In the SI system, force is expressed in newtons—see Did You Know?)

The good news is that you have obtained a force advantage, but the bad news is that you're paying for it with a distance penalty. Mechanical work is the product of force times the distance the force moves through ($W = Fd$), and this product remains constant.

In the right-side up position, the plunger pushes on the arm to raise it. But when you turn the whole assembly upside down, the syringe *pulls* on the arm to raise it, just like your muscles do with your own arms. The muscle that allows your forearm to lift things, called the *biceps*, is attached near your shoulder and just below your elbow. When the biceps contracts, it has the same effect on your arm as the syringe has on the hydraulic arm when the assembly is upside down. In both cases, a large force is exerted so that a small weight can be lifted, but the weight can be lifted a large distance compared to the distance the force moves (the distance the syringe plunger moves, or the distance your muscles contract).

So What?

Hydraulic systems are used in countless applications: brakes and steering on cars; hydraulic lifts and jacks for

The small hydraulic piston shown in this photo can lift and dump tons of earth from this truck.

servicing cars; airplane wing flaps, stabilizer controls, and landing gear; mechanical arms on garbage trucks; blades on bulldozers; and so on.

Did You Know?

Under Pressure

Automobile and bicycle tire pressures in the United States are commonly expressed in pounds per square inch (abbreviated psi), a unit from the English system of measurement. In the modern metric system (Système International d'Unités, or SI), the unit of pressure is the pascal (symbol Pa). One pascal is equal to one newton per square meter (n/m^2). Because 1 psi is approximately 6900 Pa, you can see that the pascal is a very small unit. For this reason, pressures are often expressed in kilopascals (symbol kPa; 1 kPa = 1000 Pa).

A Load on Your Shoulders

The average atmospheric pressure at sea level is equal to 14.7 psi, or 101 kPa.

Going Further

Robot Arm

Design and build a hydraulic arm that has more than one motion. Figure 7 shows an example of an arm that turns on its base in addition to lifting things, and figure 8 shows a close-up of the linkage used on the base. Can you design and build an arm that also has a "wrist" and "hand" at the end?

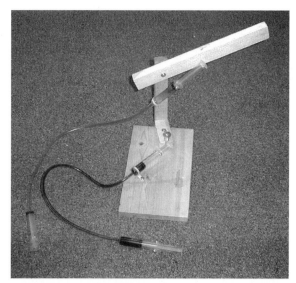

Figure 7 *Dual-action arm*

Figure 8 *Close-up of syringe and linkage for dual-action arm*

Credits & References

Modesto Tamez, Crans Squire, and Pablo Dela Cruz contributed to the development of this snack.

Bloomfield, Louis. *How Things Work: The Physics of Everyday Life.* New York: John Wiley & Sons, 1997. There is an excellent discussion of hydraulic elevators on pages 236–237.

Cameron, John, James Skofronick, and Roderick Grant. *Physics of the Body,* 2d ed. Madison, Wis.: Medical Physics Publishing, 1999. Pages 41–50 have a good discussion of the biomechanics of the arm.

Hyperbolic Slot

How do you put a straight rod through a curved slot?

If you see a straight rod and a curved slot, your common sense tells you that the only way the rod can fit through the slot is end first—but that's not quite the case. If the rod is angled and rotated through space, it describes a three-dimensional shape with a hyperbolic cross section. If the slot is the exact shape of this hyperbola, you can make the straight rod pass through it.

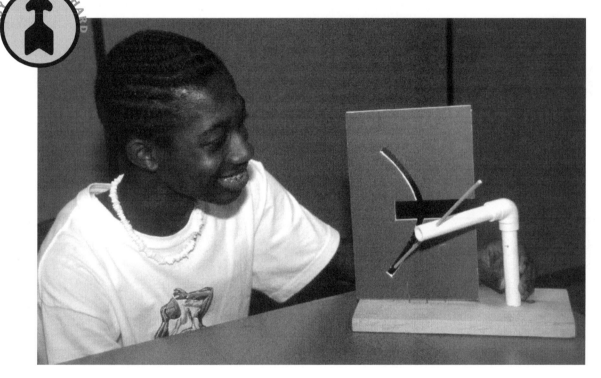

Materials

- pencil
- wood rectangle for base, approximately 7 in × 12 in (18 cm × 30 cm), can be made from $\frac{3}{4}$-in standard shelving or from plywood
- electric drill
- drill bits, $\frac{1}{8}$ in, $\frac{3}{16}$ in
- ruler
- wooden dowel, $\frac{1}{2}$-in diameter, 5 in (13 cm) long

- wooden dowel, $\frac{3}{16}$-in diameter, 11 in (28 cm) long
- screwdriver
- 1 #6 flat-head wood screw, $1\frac{1}{2}$ in long
- 2 pieces of PVC pipe, $\frac{1}{2}$ in, Schedule 40, 5 in (12 cm) long
- rubber band (optional)
- glue (optional)

- 1 PVC 90° elbow, $\frac{1}{2}$ in
- template (see figure 5)
- foam core, 7 in × 11 in (18 cm × 28 cm); cardboard can be used, but tends to deform more easily
- X-Acto knife or utility knife
- 4 finishing nails, $1\frac{1}{2}$ in long
- hammer

Figure 1

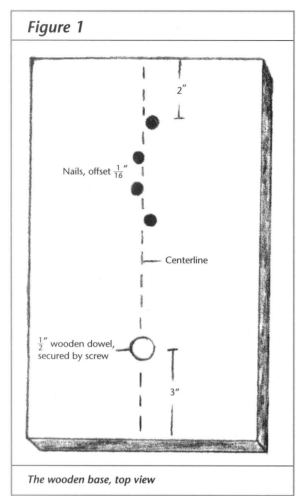

Nails, offset $\frac{1}{16}$"

2"

Centerline

$\frac{1}{2}$" wooden dowel, secured by screw

3"

The wooden base, top view

Figure 2

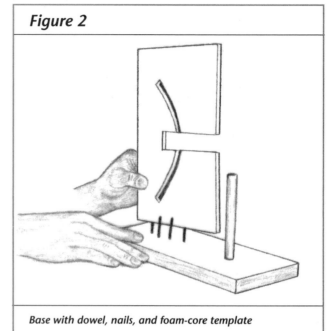

Base with dowel, nails, and foam-core template

1 Draw a line lengthwise down the center of the board that will be the base.

2 Drill a $\frac{1}{8}$-inch-diameter hole all the way through the base, 3 inches (7.5 cm) from one end and on the centerline.

3 Drill a $\frac{1}{8}$-inch-diameter hole about $\frac{3}{4}$ inch deep in the center of one end of the $\frac{1}{2}$-inch-diameter dowel. Drill this hole as straight into the center of the dowel as possible.

4 Screw the $1\frac{1}{2}$-inch screw through the hole in the base, from the bottom side of the base to the top side. (The flat head of the screw should be flush with the bottom side of the base, and the pointed end of the screw should stick straight up through the top of the base). Place the hole in the end of the $\frac{1}{2}$-inch-diameter dowel on the point of the protruding screw, and screw the dowel onto the screw until the end of the dowel is tight against the base (see figures 1 and 2).

5 Photocopy the slot template (see figure 5) at the end of this snack. Lay the photocopy on the foam core. Use this template as a guide to cut the slot in the foam core with an X-Acto knife or utility knife. Be sure the knife you use is very sharp, because foam core is difficult to cut neatly with a dull knife.

6 Make a small mark along the centerline of the base, 2 inches (5 cm) from the edge of the base far- thest away from the rotating PVC assembly (see figure 1). Make additional marks at 3 inches, 4 inches, and 5 inches (7.5 cm, 10 cm, and 12.5 cm) from the edge.

7 About $\frac{1}{16}$ inch (2 mm) to one side of the 2-inch and 5-inch marks, hammer finishing nails into the base so that they are firmly in place but stick up about an inch (2.5 cm) above the base. Then hammer in nails about $\frac{1}{16}$ in (2 mm) to the other side of the 3-inch and 4-inch marks. This will make a holder for the foam core (see figures 1 and 2).

8 Wedge the bottom edge of the foam core between the two sets of nails so that the core is held in a vertical position with its bottom edge resting on the centerline. You may have to bend the tops of the nails slightly inward or outward to adjust the pressure on the foam core so that it is held firmly in place yet can still be moved to adjust its position (see figure 2).

9 Drill a $\frac{3}{16}$-inch hole all the way through the diameter of one of the $\frac{1}{2}$-inch PVC pipes, 1 inch (2.5 cm) from one end. Insert the $\frac{3}{16}$-inch wooden dowel through the holes until it is centered. (If the dowel doesn't remain in place, wrap a rubber band around the upper section of the dowel and then slide the rubber band down until it is up against the PVC at the hole, or use a little glue to hold the dowel in place.) Attach the other end of this PVC pipe to one end of the 90-degree PVC elbow. Attach the second PVC pipe to the other end of the 90-degree elbow (see figure 3).

Figure 3

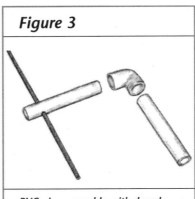

PVC pipe assembly with dowel

10 Slip the open end of the second PVC pipe over the $\frac{1}{2}$-inch dowel protruding from the base (see figure 4). Use the $\frac{3}{16}$-inch dowel as a handle to gently rotate the upper PVC pipe until the $\frac{3}{16}$-inch dowel is close to a 45-degree angle with the horizontal.

Figure 4

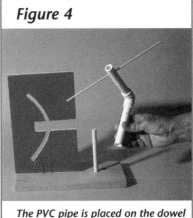

The PVC pipe is placed on the dowel so it will turn.

11 Adjust the position of the foam core and the orientation of the $\frac{3}{16}$-inch-diameter dowel arm to allow the dowel arm to pass smoothly through the slot as the arm assembly is rotated. You may have to modify the size of the slot or cut some material off the bottom of the foam core.

Caution: Hyperbolic slot ahead!

To Do and Notice

Rotate the arm assembly, and observe the straight rod passing smoothly through the hyperbolic slot. Try to focus your attention on the shape that the rod sweeps out in space as it rotates.

What's Going On?

As the rod swings around, it sweeps out a three-dimensional shape called a *hyperboloid*. A two-dimensional cross section of the hyperboloid is a shape called a *hyperbola*.

The rod passes through the slot cut into the foam core because the slot has the same hyperbolic shape as the cross section of the hyperboloid made by the rotating rod.

So What?

Being able to visualize how the rod goes through the curved slot may help you the next time you have to move a long sofa through a doorway, along a hallway that turns a corner, or up a spiral staircase!

Did You Know?

A hyperbola can be described by a specific mathematical equation:

$$y = \frac{1}{x}$$

Going Further

No Template

If you want to learn more about hyperbolas, you can construct the hyperbolic slot without using the prepared template. Hold a file folder perpendicular to the base, along the centerline of the base between the dowel and the far end of the base. Plot the hyperbolic curve by making marks where the ends of the rotating arm cross the plane of the file folder, and where the midpoint crosses the plane. Bend a piece of flexible material so that it crosses each of the three points and makes a smooth curve. Cut out this curve from the folder. Place the folder between the nails on the base, and rotate the arm slowly through the slot. Make additional cuts where necessary to allow the wooden arm to move smoothly through the slot without touching it. Then either use the folder itself for the slot (if it is stiff enough) or transfer the slot pattern to foam core and cut it out with an appropriate tool.

No Slot

Change the angle of the dowel to vertical and visualize the shape of a slot that would allow the dowel to pass through it. Then position the dowel horizontally and visualize the shape of a slot it would need now.

Credits

This snack is based on the Exploratorium exhibit of the same name.

Figure 5 *Hyperbolic slot template*

Light Conversation

The talk is simple, but illuminating.

An automatic night-light has a built-in photocell sensor with circuitry that causes the bulb to go on when the light striking the photocell drops below a certain level. A single automatic night-light works well for illuminating a dark hallway, but if you have two of them on extension cords you can have fun making them "communicate" with each other.

Materials

- 2 night-lights with photocell light sensors (available in drugstores, discount stores, home improvement stores, and hardware stores)
- 2 extension cords, 6 ft (2 m) or longer

- 2 electrical outlets within about 6 ft (2 m) of each other (If you don't have two nearby outlets, use a power strip or a third extension cord with at least two plug-ins on its end.)

- mirror (small, handheld)
- credit card (or any opaque card about the same size)

ASSEMBLY

1 If possible, remove the plastic diffuser shade from each light so that you can see the bare bulbs, as shown in figure 1. (**CAUTION:** Don't remove the shades if it seems that you will break any parts of the night-lights in the process.)

Figure 1

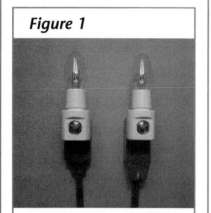

Night-lights with photocell light sensors

2 Plug each light into a different extension cord and plug both cords into electrical outlets.

To Do and Notice

Darken the room. One or both of the night-lights should turn on.

Experiment with the night-lights by moving them around. Watch the behavior of the lights, and use it to predict when they will go on and off.

Unplug one of the lights and put it aside. Turn the room lights on. What happens to the light that is still plugged in? Completely cover the photocell sensor (the small window below the bulb on the base of the light) with your hand. What does the light do?

Darken the room. The night-light that is still plugged in should go on. Hold the mirror so that the light from this night-light shines into its own photocell. Adjust the position of the mirror until you get the light to flicker. Can you explain what's happening?

Keep the room dark and plug in the second night-light again. Place the lights on a table so that each bulb shines into the other light's photocell, as shown in figure 2a. Initially, both lights may go out for an instant, but one or the other should come back on. Then slowly slide an opaque object such as a credit card between the bulb that is *unlit* and the photocell opposite it (figure 2b), and continue sliding it so that it passes between the bulb that was initially lit and the photocell opposite it (figure 2c). Keep sliding until the card is no longer between the units (figure 2d).

Slowly slide the card back between the units in the opposite direction. What happens?

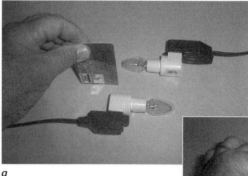

a

Figure 2 *Sliding an opaque object between the night-lights (These photos were taken with the room lights on and don't show how the night-lights turn on and off.)*

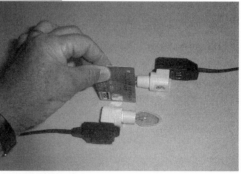

b

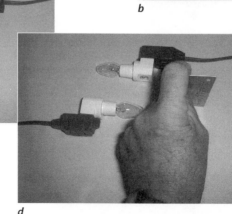

c

d

What's Going On?

By positioning the two night-lights in certain ways, you are causing them to affect each other's behavior, and when you use the mirror, you are causing a single night-light to affect its own behavior.

When light from the bulb of one unit shines on the photocell sensor of the second unit, the bulb on the second unit turns off. If the second unit is moved so that the light no longer hits the photocell, it lights up again. And if the light from the second unit hits the photocell of the first unit, the bulb in the first unit will turn off. Each night-light, in a sense, is telling the other what to do.

When you use the mirror to reflect light from a night-light back into its own photocell, its output (light) becomes its input, and the unit begins communicating with itself. Light reflected to the photocell causes the bulb to turn off; then, when light is no longer reflected, the bulb turns back on. There is a slight time delay in the cooling and heating of the filament in the bulb, so this on–off–on–off sequence is seen as flickering.

When you slide the credit card past the unlit bulb, no change occurs, but when the card goes past the lit bulb, its light is blocked and no longer falls on the photocell opposite it. This turns on the bulb that was originally unlit, and its light falls on the photocell across from it, turning off the bulb that was originally lit.

These interactions occur because each night-light is a simple logic device: Its output (light or no light) depends on the input it receives (light or darkness). The circuitry in the unit carries out the simple rules summarized in the following table:

Input	Output
Light (1)	Off (0)
Darkness (0)	On (1)

So What?

When part of the output of a device is fed back into the device to change its behavior, a feedback loop is created. When you use the mirror to make the night-light flicker, you are making a feedback loop. If the feedback increases the output of the device, it is called *positive feedback;* if it decreases the output, it is *negative feedback.*

With the flickering night-light, the output from the bulb works as negative feedback because when it is positive (on) it causes a negative result (the light going off).

Some other devices that use negative feedback are the cruise control on a car, the automatic pilot on an airplane, the thermostat on a heating system, and a weather vane. An example of positive feedback is the loud screech heard when sound from the speakers in a public address system gets fed back into the microphone.

The alternating on–off pattern obtained by sliding the credit card between the two night-light units might be thought of as a crude analogy to what is known in electronics as a flip-flop circuit, or bistable circuit.

Such a circuit has two stable states and two inputs, and switches back and forth between the two—or flip-flops—each time a pulse is applied to the other input. One application of these circuits is the counting of binary digits (on–off, or 0–1) in computing systems.

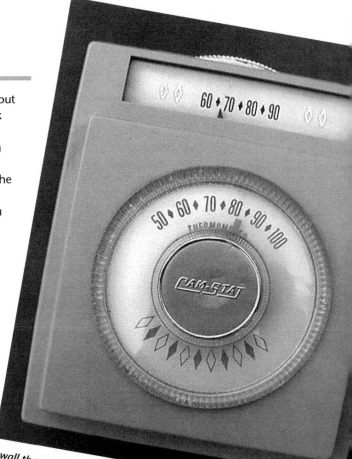

A wall thermostat monitors the temperature in a building, turning on heat or cooling air as needed to maintain a constant preset temperature.

Going Further

Night-Light Conference

Obtain several night-lights, and have them "talk to each other" in a dark room. Try different locations and orientations of the lights, and see what happens. This experiment is best carried out by having a different person hold and manipulate each light, or by mounting the lights on a table or board in such a way that their positions and orientations can be easily adjusted.

Which Is Brighter?

Think of a way to use the pair of night-lights to investigate the relative brightness of different bulbs. Try comparing a clear bulb, such as the one that comes with the night-light, with a white frosted Christmas tree light, or even with colored Christmas tree lights. Can you think of a way to make your investigation quantitative?

Test Your Logic

Predict what will happen if you repeat the last two steps of the To Do and Notice section, but this time start by sliding the credit card between the bulb that is lit and the sensor opposite it.

Membrane Aerophone

Make a simple pseudo-saxophone.

Using a balloon, a film canister, a PVC tube, and a straw, you can make a musical instrument that sounds surprisingly like a saxophone. When you blow into the instrument, the stretched balloon acts as a vibrating membrane to produce the sound.

Materials

- clamp (suitable for securing items to be drilled)
- film canister (black plastic ones work better than clear plastic because the plastic is thinner)
- electric drill
- $\frac{3}{4}$-in spade bit (the kind with small cutting tips on the outer edges as shown in figure 1 works well)
- $\frac{1}{4}$-in drill bit
- scissors
- drinking straw with flexible end
- ruler or tape measure
- hot glue gun and glue sticks
- $\frac{1}{2}$-in diameter PVC pipe, 10 to 12 in (25 to 30 cm) long
- pencil
- sandpaper (small piece)
- balloon (9-in round works well)
- rubber band (optional)

Caution: Assembly requires drilling into a film canister and PVC pipe. Have these items clamped safely in place when you drill.

1 Drill a hole in the bottom of a film canister using a $\frac{3}{4}$-inch spade bit (see figure 1).

2 Drill a hole in the side of the film canister, near the bottom, using a $\frac{1}{4}$-inch bit. This hole is for inserting the straw (see figure 2).

Alternative Construction

1. You can use a $\frac{13}{16}$-inch spade bit, which will make a slightly larger hole than the $\frac{3}{4}$-inch bit—then you can wrap tape around the pipe if the hole is too large.

2. It's possible (although time-consuming) to drill the hole with the spade bit by hand.

Figure 3

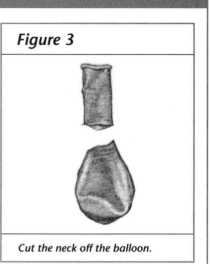

Cut the neck off the balloon.

6 Place the remaining part of the balloon over the open end of the film canister and pull it tight like a drum head. Then use the part of the balloon neck that you cut off as a rubber band to hold the stretched balloon membrane in place as shown in figure 4. (You can use an actual rubber band if it's more convenient.)

Figure 1

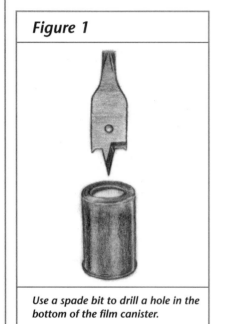

Use a spade bit to drill a hole in the bottom of the film canister.

Figure 2

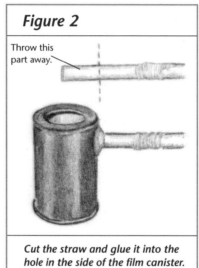

Throw this part away.

Cut the straw and glue it into the hole in the side of the film canister.

3 Cut off the end of the straw nearest the flexible bend so that about 1 inch (2.5 cm) remains past the bend. Insert this end of the straw a very short distance into the hole in the side of the film canister. (Later, when the PVC pipe is inserted into the bottom hole, it should not touch the straw.) Then hot glue the straw in place as shown in figure 2. Be careful not to touch the straw with the metal tip of the glue gun because that could melt the straw.

4 Place a ruler along the length of the PVC pipe. Make small *x* marks 2 inches, 4 inches, and 6 inches (5 cm, 10 cm, and 15 cm) from one end of the pipe. Drill a $\frac{1}{4}$-inch hole at each *x* mark. Sand around each hole to remove any significant burrs or sharp edges; do the same for the ends of the PVC pipe.

5 Cut off the first inch of the neck of the balloon as shown in figure 3. Save the piece you cut off, because you will use it as a rubber band.

Figure 4

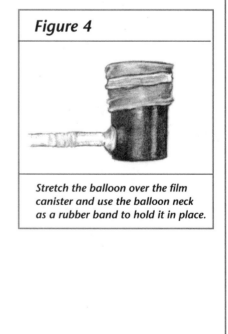

Stretch the balloon over the film canister and use the balloon neck as a rubber band to hold it in place.

Figure 5

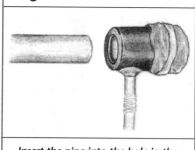

Insert the pipe into the hole in the bottom of the film canister.

Figure 6

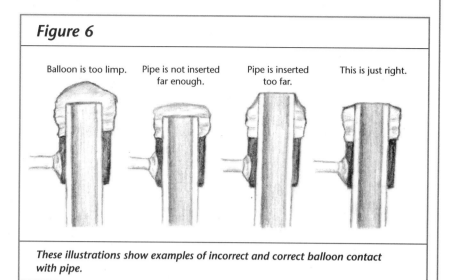

Balloon is too limp. Pipe is not inserted far enough. Pipe is inserted too far. This is just right.

These illustrations show examples of incorrect and correct balloon contact with pipe.

7 Insert the end of the pipe farthest from the holes into the film canister bottom (see figure 5). If you have difficulty getting the pipe into the hole, try twisting the pipe as you insert it. With a little patience and perseverance, it should eventually slip into the hole. Once in a great while, the edge of the hole on the film canister may actually split or tear; if this happens, you will have to start over with a new film canister.

8 Adjust the PVC pipe so that its end just barely contacts the stretched balloon and you can see the outline of the pipe on the balloon (see figure 6).

To Do and Notice

Blow on the straw, and listen to the sound. If the instrument doesn't produce a saxophone-like sound with relatively little effort, try adjusting the PVC pipe so that it contacts the stretched balloon a little more tightly or loosely. Also, check to be sure that the balloon membrane has some tension in it and is not just limp.

Cover all three of the holes in the pipe with your fingers as you blow. Note the difference in pitch compared to leaving the holes open. Try covering just one or two holes. How many different notes can you make?

The pitch of the sound produced depends on breath pressure and membrane tension. Membrane tension can be adjusted by stretching the membrane tighter and adjusting the rubber band while holding the membrane in place or by touching the membrane while playing.

What's Going On?

When you blow through the straw, pressure builds up in the space between the outside wall of the pipe and the inside wall of the film canister. This

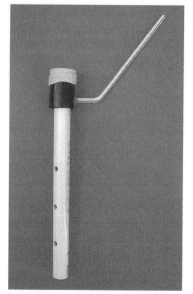

Figure 7 *Here's what a completed aerophone looks like.*

pressure lifts the stretched balloon membrane, allowing air to be released into the PVC pipe. The balloon then springs back into place, sealing the pipe again, and the whole process is repeated. All this happens very rapidly, many times each second.

The frequency, or pitch, of the sound you hear when you blow into the instrument depends in part on the rate at which the membrane bounces back and forth on the end of the pipe. (The vibration of the balloon membrane is actually very complex, and only certain patterns of vibration resonate with the tube to produce the sounds that you hear.)

The length of the pipe is important in determining which sound waves are resonant and, therefore, which frequencies are heard. Opening or covering the holes in the pipe produces different notes because it changes the length of the pipe. The effective length of the pipe is from the top to the first open hole or to the bottom of the pipe if all holes are covered.

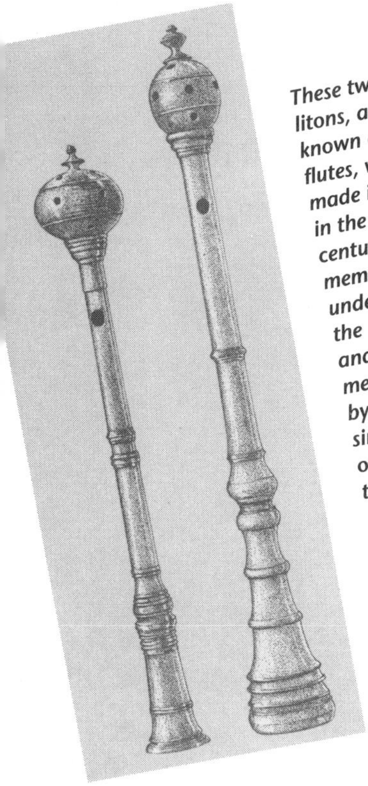

These two mirlitons, also known as onion flutes, were made in Europe in the eighteenth century. The membrane is under the cap on the upper end, and the instrument is played by speaking or singing into one of the holes in the cap.

So What?

The membrane aerophone is easy to play. It's a good instrument to build for elementary school students.

Did You Know?

Meet the Mirlitons

The membrane aerophone is actually a member of a family of musical instruments known as mirlitons, which includes the familiar kazoo. The sound of the kazoo depends on the player's voice, however, while the membrane aerophone in this snack is more closely related to mirlitons that are played by blowing.

Going Further

Do Re Mi

Experiment with the number, size, and spacing of the holes you drill in the pipe to create an instrument on which you can play a musical scale.

Trombone Aerophone

Get a section of $\frac{3}{4}$-inch thin-wall PVC pipe to use as a slide for your aerophone. Wrap masking tape around the end of the aerophone to make a gasket that allows the wider piece of pipe to slide freely onto the aerophone but which forms an airtight seal. Now cover the holes in the aerophone with tape and use the slide to vary the length—and therefore the pitch—of the instrument.

Crank It Up

Cut the top off a plastic soda bottle and attach it to the end of the aerophone as shown in figure 9. The plastic bottle allows a more efficient transfer of sound energy to the air, thus amplifying the sound.

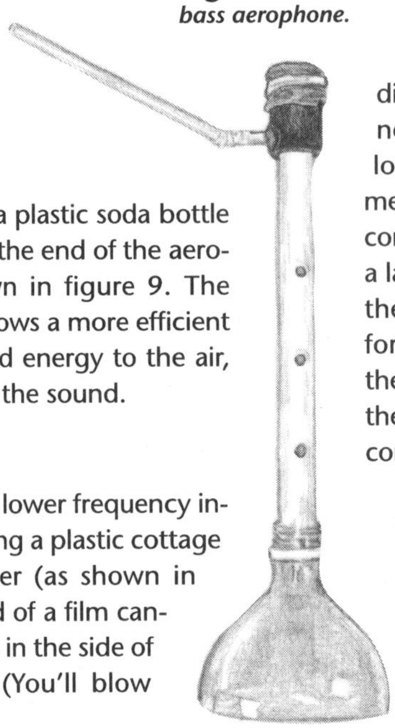

More Bass-ic

You can make a lower frequency instrument by using a plastic cottage cheese container (as shown in figure 8) instead of a film canister. Drill a hole in the side of the container. (You'll blow

Figure 8 *Use a cottage cheese container to make a bass aerophone.*

directly into the hole; no straw is necessary.) Instead of the balloon, use plastic wrap as the membrane, and secure it over the container with a rubber band. Use a larger diameter piece of PVC for the tube. Use modeling clay to form a reasonable seal between the tube and the hole you make in the bottom of the cottage cheese container.

Figure 9 *Amplify your aerophone using a plastic soda bottle.*

Credits & References

This snack was developed by Fran Holland of the Exploratorium.

Botermans, Jack, Herman Dewit, and Hans Goddefroy. *Making and Playing Musical Instruments.* Seattle: University of Washington Press, 1989. Includes many photographs of nicely crafted homemade musical instruments from different regions of the world, made with simple materials. There's a section about membranophones on pages 70–83.

Diagram Group. *Musical Instruments of the World.* New York: Facts on File, 1976. This book is subtitled "An illustrated encyclopedia with more than 4000 original drawings," and it is a truly comprehensive and fascinating collection of information and illustrations.

Hopkin, Bart. *Making Simple Musical Instruments.* Asheville, N.C.: Lark Books, 1995. There are instructions for building and playing approximately 30 musical instruments, including many made from common materials. Excellent photos and drawings accompany the text.

Hopkin, Bart. *Musical Instrument Design.* Tucson, Ariz.: See Sharp Press, 1996. An invaluable reference for designing and experimenting with homemade musical instruments. Chapter 6 covers aerophones, and Chapter 7 describes membranophones.

Modulated Coil

Using a simple electromagnet, you can wirelessly transfer the sound from a radio to the speaker of a tape player.

Hear the magnet!

Materials

- wire stripper or knife
- about 3 ft (1 m) of insulated wire (e.g., RadioShack #20 or #22 solid copper wire with plastic insulation)
- steel bolt, about $\frac{1}{4}$-in diameter and 2 in long (nut optional); exact size of bolt is not critical
- audio cable, 6 ft (2 m), $\frac{1}{8}$-in phone plug on one end and two alligator clips on the other (e.g., RadioShack #42-2421; a phone plug is sometimes called a mini plug)
- small radio with headphone jack (e.g., RadioShack #12-799)
- portable tape cassette player with speaker (if the player doesn't have its own speaker, you'll have to have the headphones on)

1 Use a wire stripper or knife to remove about half an inch (1.2 cm) of the plastic insulation from each end of the wire. (If you happen to have enamel-insulated wire instead of plastic-insulated wire, use sandpaper to remove the enamel.)

2 If you have a nut for the bolt, screw it onto the end of the bolt. It may help keep the wire that you're about to wrap onto the bolt in place, but it isn't essential.

3 Start wrapping the wire around the bolt, leaving about an inch (2.5 cm) of wire free on the starting

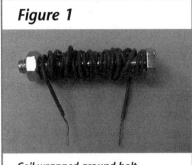

Figure 1

Coil wrapped around bolt

end of the wire. Begin as close as you can to one end of the bolt and proceed toward the other end. When you reach the other end, start another layer and proceed back toward

the original end, but keep wrapping in the same direction (i.e., clockwise or counterclockwise, whichever direction you began with; see figure 1). If you reverse the direction of your wraps, you'll cancel the effect of the wire you wrapped initially. Keep wrapping the wire around the bolt, building up multiple layers if necessary, until you have at least 20 wraps. When you've finished wrapping, leave another inch (2.5 cm) of wire free.

4 Attach the two alligator clips on the audio cable to the ends of the wire on the bolt.

To Do and Notice

Turn on the radio and find a radio station with a strong, clear signal. Adjust the volume to medium-high. Plug the phone plug on the audio cable into the headphone jack on the radio. When you do this, you will no longer hear the radio, since the signal is being fed to the headphone circuit instead of to the speaker.

Be sure there is no tape in the tape player, and then press the play button. Adjust the volume control on the tape player to medium-high. Since there is no tape in the player, you should not hear any significant sound.

Bring the wire-wrapped bolt near the head of the tape player. You should hear the sound from the radio station playing through the speaker of the tape player. (Remember: If the tape player doesn't have its own speaker, you'll need to have the headphones on.)

What's Going On?

The radio sends an electric current through the audio cable and through

the coils of wire wrapped around the bolt. The wire-wrapped bolt becomes an electromagnet, with the strength of its magnetic field determined in part by the size of the current flowing through the coils. Because the current carries an audio signal, it varies in strength, causing the magnetic field of the electromagnet to vary also.

The head of the tape player is essentially a device for detecting very small variations in a magnetic field. Normally it detects variations in the magnetic field on the audiotape as the tape travels by. In this case, however, it senses the fluctuating magnetic field in the coils of wire wrapped around the bolt.

So What?

The "T" mode of a hearing aid, which is designed to be used with a telephone, works on the principle of magnetic field coupling demonstrated by this snack. A telephone has a magnet whose field varies with the oscillations of the sound signal. A hearing aid, like the head of the tape recorder in the snack, detects small variations of the magnetic field. This fluctuating mag-

netic field induces current in the pickup coils of the hearing aid, and the current is converted to sound. This "T" mode eliminates the annoying high-pitched audio feedback to the hearing-aid microphone that is often present and can be made worse by covering the hearing aid with the telephone headset.

Going Further

Iron Versus Air

In principle you could use the coil of wire alone, without the bolt. You would then have an electromagnet with an air core rather than an iron core. The iron core, however, greatly intensifies the magnetic field. What would you have to do to achieve the same effect with an air core? Check your reasoning by building an air-core electromagnet.

References

Rathjen, Don. "Trick of the Trade: Modulated Coil," *The Physics Teacher*, Vol. 36, No. 7, October 1998, p. 416.

Modulated LED

Listen to a beam of light.

EASY MEDIUM HARD

Audio signals can be carried in radio waves through space and in electrical pulses through wires. Other forms of electromagnetic radiation, including visible light, can carry audio signals, too. You can build a simple device in which the signal from a radio is transmitted on a beam of light traveling between a light-emitting diode (LED) and a solar cell.

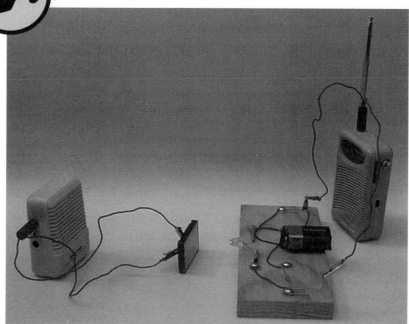

Materials

- 7 alligator clips (e.g., RadioShack #270-380 mini alligator clips)
- needle-nose pliers
- insulated copper wire, 22- or 20-gauge solid copper, 5 ft (1.5 m)
- wire stripper
- wooden board for base, approximately $3\frac{1}{2}$ in × 8 in (9 cm × 20 cm), can be made from $\frac{3}{4}$-in standard shelving or plywood
- 470-ohm resistor (RadioShack #271-1317)
- 2 paper clips
- 9-volt battery
- Velcro with adhesive back, approximately 1 in × 1 in (2.5 cm × 2.5 cm), to hold the battery to the board

- 9-volt battery snap connector (e.g., RadioShack #270-325 or #270-324)
- 6 sheet-metal screws, #8 × $\frac{5}{8}$ in
- 6 small flat steel washers (SAE #10)
- light-emitting diode (LED) (e.g., RadioShack #276-066 High Brightness Red LED)
- 2 phone plugs, $\frac{1}{8}$ in (e.g., RadioShack #274-286 or #274-287, sometimes called mini plugs; see Helpful Hint on page 52)
- amplified speaker (e.g., RadioShack #277-1008)
- solar cell (e.g., Edmund Scientific Co., 800-728-6999, www.edsci.com, #30398-09)

- small radio with headphone jack (e.g., RadioShack #12-799)

NOTE: Two premade 6-ft audio cables with a $\frac{1}{8}$-in phone plug on one end and two alligator clips on the other end (e.g., RadioShack #42-2421) can be substituted for the two phone plugs, about 3 ft (1 m) of the wire, and four of the alligator clips in the list of materials above. One of the cables will have to be altered slightly so that its alligator clips are far enough apart to be connected to the paper clips at screws 1 and 4 (see figure 1 on next page).

Figure 1

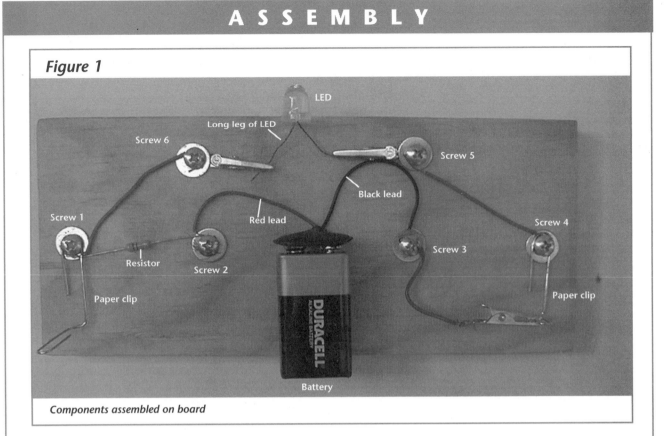

LED

Long leg of LED

Screw 6

Screw 5

Black lead

Screw 1

Red lead

Screw 4

Resistor

Screw 2

Screw 3

Paper clip

Paper clip

DURACELL

Battery

Components assembled on board

1 Flatten the ends of two of the alligator clips by bending the tabs outward with the needle-nose pliers (see figure 2).

Figure 2

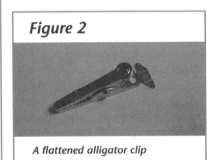

A flattened alligator clip

2 Cut the wire into three 3-inch-long (8-cm) pieces and four 12-inch-long (30-cm) pieces. Strip about a half inch (1.2 cm) of insulation off all the ends.

3 Attach one of the short wires to an alligator clip.

4 Assemble components on the board as shown in figure 1. The ends of the short wires, the ends of the resistor, the paper clips, and the battery leads are all held in place between a screw head and its washer. The alligator clips with the flattened ends are held in place under their washers (at screws 5 and 6). Note the following:

a. The red lead from the battery should be attached to screw 2, and the black lead to screw 3.

b. The long leg of the LED should be attached to the jaws of the alligator clip at screw 6 and the shorter leg to the jaws of the alligator clip at screw 5. The LED will allow current to flow in only one direction, so it's important to connect the leads properly. The longer lead from the LED is the positive lead.

c. The alligator clip on the wire coming from screw 3 is used as a switch; when it's connected to the paper clip and the battery is in place the LED should be lit.

➡ Helpful Hint

Be sure you have a mono plug. A mono plug can only be wired one way, and it will work with either a mono or stereo radio. A stereo plug can be wired in different ways, some of which may not work with particular radios (see figure 3 and step 6).

Figure 3

A mono plug (right) has one black band. A stereo plug (left) has two.

Figure 4

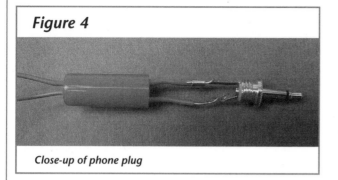

Close-up of phone plug

Figure 5

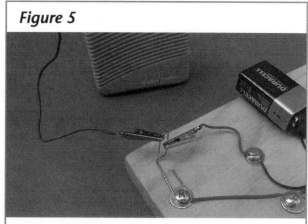

Attach the alligator clips from the radio to the paper clips on the board.

5 Note where the battery lies on the board, and use the Velcro to hold it in place.

6 (**NOTE:** Skip this step if you are using the premade audio cables described in the Materials section.) Unscrew the plastic cover from one of the phone plugs (see Helpful Hint on page 52). Attach a 12-inch (30-cm) wire to the center terminal of the plug, and another 12-inch (30-cm) wire to the outer terminal (see figure 4). Slide the plastic plug cover onto the two wires and screw it back onto the plug. Be sure that the wire ends at the terminals don't touch each other—you may want to wrap a little tape around one of them to prevent contact. Attach an alligator clip to the other end of each of the wires. Repeat this process with the other phone plug, using two more wires, and two more alligator clips.

7 Insert one of the phone plugs into the input jack on the amplified speaker. Attach the two alligator clips to the contacts on the solar cell. Turn on the speaker and hold the solar cell toward the light. You should hear static from the speaker. Turn the speaker off while you finish the assembly process.

8 Turn on the radio and find a loud, clear signal. Attach the alligator clips of the second phone plug assembly to the paper clips on the board (see figure 5). It doesn't matter which alligator clip is connected to which paper clip. Plug the phone plug into the earphone jack on the radio. Once you insert the plug into the jack, you won't be able to hear the radio anymore. You may now be able to see the LED flickering, although it may fluctuate so quickly that you perceive it as a steady light.

➡ Helpful Hint

To test the solar cell and amplifier, turn on the amplifier and hold the solar cell in the light from a fluorescent bulb. If the system is working, you should hear a loud hum. Most fluorescent bulbs flicker on and off 120 times a second; the hum is produced by this fluctuation. (Some modern high-frequency fluorescent bulbs flicker on and off much faster and will not produce a hum.)

You can also test the solar cell by moving a comb back and forth between a small light source such as a glowing LED and the solar cell. Because of the spaces between the teeth, the comb alternately blocks the light and lets it pass. This creates pulses of electric current in the solar cell that, when passed through the amplified speaker, produces a "sawing" sound.

To Do and Notice

Hold the solar cell about 12 inches (30 cm) from the LED and point it directly at the LED. Turn on the amplified speaker again. You should now be able to clearly hear the radio signal coming from the amplified speaker. If you have difficulty, or if there seems to be significant static, block peripheral light from hitting the solar cell, or turn off the room lights if possible.

Put your hand or a piece of paper between the LED and the solar cell. The radio signal should stop.

Try reversing the connections on the LED, attaching the long leg to screw 5 and the short leg to screw 6. You will find that the LED won't light.

What's Going On?

The battery provides a steady DC current to light the LED. Under the influence of the battery alone, the LED glows with a fixed brightness. The resistor limits the current so the LED does not burn out.

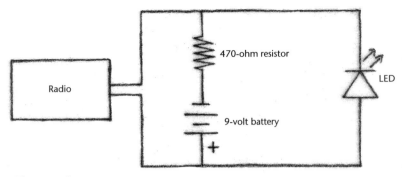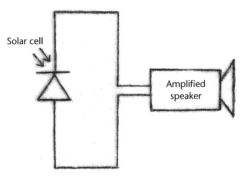

Figure 6 *Circuit diagram with components labeled*

When the radio is turned on, the weak but fluctuating radio signal is added to the constant signal from the battery. The LED still glows, but now it flickers in synchronization with the radio signal, as the amount of current passing through it varies.

The flickering light hitting the solar cell causes it to generate electrical signals which again vary in synchronization with the original radio signal. These signals are amplified and fed to the speaker, recreating the original sounds from the radio. (The preceding process is represented in figure 6.)

Placing an opaque barrier between the LED and the solar cell cuts off the light, and the solar cell is unable to generate any electrical signals.

So What?

Modern communications systems, such as long-distance phone lines and high-bandwidth communication lines for computers, commonly use modulated signals carried on a beam of light. The relatively high frequencies of visual light can carry a lot more information than lower frequency radio waves. The light that is used is normally from a laser, not an LED, and the signals are carried by a clear fiber-optic cable, rather than through air, but the principle is the same.

Going Further

Which LED Works Best?

Try several different LEDs and find out which results in the clearest sound or transmits the sound farthest. Does the relative brightness of an LED affect the results?

Cancel the Noise

It can be hard to use your system in a room lit by fluorescent lights because the lights flicker on and off, producing an annoying hum. To get rid of the hum, you can make a noise-canceling receiver with two identical solar cells.

Hook the solar cells together in a series, attaching the positive side of one to the positive side of the other. Attach the amplifier to the two negative terminals.

Shine the light from the LED on only one of the solar cells while the room light shines on both. The signal from the LED is converted to electricity by one of the cells, and the signal from the room light is converted into electricity by both of the cells. Because the cells are hooked together so that they oppose each other, the two signals from the room light cancel each other. This allows you to hear the radio signal in a noisy environment.

Credits & References

This snack is based on the Exploratorium exhibit Modulated Laser.

Macaulay, David. *The Way Things Work*. Boston: Houghton Mifflin, 1988.

Mims, Forrest M. III. *Getting Started in Electronics*. RadioShack, 2000. (RadioShack #62-5004)

In World War II, navy signalmen used flashing lights to send messages between ships.

Oil-Spot Photometer

Greased lighting.

A drop of grease or oil on white card stock is easily visible because it changes the way that light interacts with the card. In this snack, you take advantage of this effect to make a kind of light meter. When an oil spot on a card is illuminated equally from both sides, it mysteriously disappears. This allows you to compare the brightness of the light sources on either side of the card.

Materials

- 1 drop of cooking oil
- white card (e.g., 4 in × 6 in or 3 in × 5 in)
- paper towel or tissue
- 2 new 60-watt frosted incandescent light bulbs (bulbs should be as identical as possible, e.g., same brand, both standard or both "soft white")
- 2 power strips
- 2 plug-in lamp sockets (see figure 1)

- extension cord(s) as necessary to allow both lights to be placed on a table and moved about 3 ft (1 m) apart
- 1 40-watt frosted incandescent light bulb (other characteristics the same as the 60-watt bulbs if possible)
- 1 75-watt frosted incandescent light bulb (other characteristics the same as the 60-watt bulbs if possible)

- pot holder or towel for handling hot bulbs
- 1 compact fluorescent light bulb that fits a standard socket and is intended to replace a 60-watt incandescent bulb
- packaging that the light bulbs came in, containing information about power (watts) and illumination (lumens)

1 Dip your finger into a drop of cooking oil, and then press the end of your finger on the middle of the white card, forming an oil spot about a half inch to 1 inch (1 or 2 cm) in diameter. Make sure the spot is visible on both sides of the card. Use a paper towel or tissue to blot excess oil from the spot, but do not smear oil anywhere else on the card.

2 Put the two 60-watt bulbs in the sockets and plug one socket into each power strip. Then place the power strips so that the centers of the bulbs are about 6 feet (2 m) apart and turn the bulbs on. If you're short of space the bulbs can be as close as 3 feet (1 m) apart, but 6 feet is better.

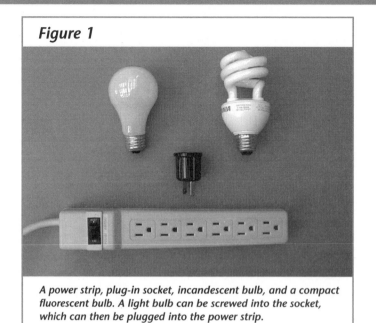

Figure 1

A power strip, plug-in socket, incandescent bulb, and a compact fluorescent bulb. A light bulb can be screwed into the socket, which can then be plugged into the power strip.

To Do and Notice

Hold the card upright between the lights, so that each side of the card directly faces one of the lights.

Move the card back and forth between the lights so that it is first closer to one light and then the other. As you do this, stand so you can keep your eye on one side of the card. What happens to the darkness and lightness of the oil spot (compared to the rest of the card) as you move the card back and forth? When is the spot darker than the rest of the card? When is it lighter?

Notice that there is a location between the lights where the grease spot comes close to disappearing. Move the card to this location, and take note of where it is relative to the two lights.

Replace the two 60-watt light bulbs with the 40-watt and 75-watt bulbs (use the pot holder or towel if the lights are hot). As before, find the position between the lights where the oil spot disappears, and notice the approximate location of the card relative to the two lights.

Replace the 40-watt and 75-watt bulbs with a 60-watt incandescent bulb and the compact fluorescent bulb that is designed to serve as a replacement for a 60-watt incandescent bulb. (**NOTE:** You may have to let the fluorescent bulb warm up for a short time to achieve full brightness.) Again, locate the position between the lights where the oil spot disappears, and notice the approximate location of the card relative to the two lights.

Hold your hand above each of the bulbs. You should easily be able to feel that the incandescent bulb is much hotter than the fluorescent bulb. Now take a look at the boxes these two bulbs came in, and notice the power (watts), illumination (lumens), and price for each bulb. (If the prices aren't printed on the boxes, see if you can find out what they are.)

What's Going On?

Compared to the normal white surface of the card, the oil spot reflects less light and transmits (lets pass through) more light. When the side of the card you are looking at is illuminated more than the other side, more light is reflecting from the rest of the card than from the oil spot, and the amount passing through the spot toward you from the dimmer side isn't enough to make up the difference. As a result, the oil spot appears darker than the rest of the card.

When the side of the card you are looking at is illuminated less than the other side of the card, there is still more light reflecting from the rest of the card than from the oil spot, but the amount passing through the spot toward you from the brighter side is more than enough to make up the difference. As a result, the oil spot appears lighter than the rest of the card.

The oil spot disappears when it is equally illuminated from both sides. In this case, the combination of the light reflected from the oil spot plus the light transmitted through it from the other side is equal to the light reflected from the card. For the two 60-watt bulbs, this point should be about halfway between the two bulbs, since they should be providing approximately equal illumination. For the 40- and 75-watt bulbs, however, the location where the spot disappears should be noticeably closer to the 40-watt bulb. The dimmer the light source, the closer you have to be to it to receive a given amount of illumination.

For the 60-watt incandescent bulb and the fluorescent "60-watt replacement" bulb, the card should again be approximately halfway between the two bulbs when the oil spot disappears, showing that the two lights are of approximately equal brightness. Fluorescent bulbs are more efficient than incandescent bulbs, converting a greater percentage of electrical energy into light rather than heat. According to the ratings on the boxes, the fluorescent bulb we experimented with uses 16 watts of electricity to give a light output of 800 lumens, whereas the incandescent bulb uses 60 watts of electricity to give a light output of 855 lumens. The fluorescent bulb gives about 90 percent of the light output of the incandescent bulb, but uses only about 25 percent as much energy to do it.

So What?

The fluorescent bulb costs more than the incandescent bulb. But, according to the manufacturer, the fluorescent will last up to ten times longer and will consume far less energy in the process. Overall, there is a significant savings in both cost and energy consumption. Replacing some household incandes-cent bulbs with fluorescent bulbs is an easy way for people to conserve energy.

Did You Know?

Orange Hot
Both the sun and most of the light bulbs we use are incandescent pro-ducers of light. That is, they emit light because they're hot. When an electric current passes through the tungsten filament of an incandescent bulb, it heats up to about 2900 Kelvin (4700°F). That's about half the tem-perature of the sun. The glowing fila-ment looks orange because it's cooler than the sun, and so it gives off pro-portionately more long-wavelength radiation, which we see as reddish-orange light. If it could get hotter, it would look more white.

Blue + Red = White
When electricity passes through a fluo-rescent bulb, the moving electrons ex-cite the mercury vapor that fills the bulb. In response, the mercury vapor emits mostly high-frequency blue light (which we can see) and ultra-violet light (which we can't see). A phosphorescent material that coats the inside of the bulb absorbs this ul-traviolet light and emits lower-frequency visible light.

Try waving a white-barreled pen over a piece of black paper in a room lit by normal fluorescent lights. You should be able to see bluish and red-dish bands as the pen waves back and forth. Why? Fluorescent lights flash on and off 120 times a second. During the moment the light is on, you see bluish light emitted by the mercury vapor. Then, a fraction of a second later when the light is off, you see red-dish light emitted by the glowing phosphors. Together, the bluish light and reddish light look white.

A Bunsen What?
A slightly more sophisticated form of the oil spot photometer you make in this snack is officially known as a Bunsen photometer.

Going Further

Double Bright
If you hold the oil-spot card between a single 60-watt bulb on one side and two adjacent 60-watt bulbs on the other, where will the "equal point" be? Try predicting the location of the equal point using this hint: Brightness varies inversely with the square of the distance from the light source. Then set up the bulbs this way and test your prediction. (See Box o' Math for an explanation.)

A Bright Idea
How much brighter is the 75-watt bulb than the 40-watt bulb? Use your photometer and the inverse square relationship cited above to figure this out.

How Much Dimmer?
Use your photometer to test the rela-tive light output of bulbs that have the same wattage but differ in some other way (e.g., a new bulb and a bulb that's been used for awhile, or two different brands or shapes of bulb, or a stan-dard bulb and a "soft white" bulb).

Light and Power
On the bulb packages, light output is given in lumens. In incandescent bulbs, are lumens and watts propor-tional? Would twice the wattage give twice the light output? How are lu-mens and watts related? Look up lumen and the related units candela and lux.

Box o' Math
Inverse Square Law

The intensity of a light (I), which the human eye sees as brightness, is the light power (P) per unit area (A):

$$I = \frac{P}{A}$$

As light moves outward from a bulb, the power spreads over a spherical area of radius (r) that increases as the square of the distance from the bulb. The area (A) of a sphere with radius r is

$$A = 4\pi r^2$$

So for a bulb of constant power (P), the intensity is

$$I = \frac{P}{A} = \frac{P}{4\pi r^2}$$

Because r^2 is in the denominator of the fraction, it's verbally described as an *inverse square*.

When the oil-spot card held between two light sources reaches the point of equal brightness, the intensities of the two lights are the same. Therefore, $I_1 = I_2$ since the power of one light is P_1 and the power of the other is P_2, while the distance from the center of one light to the card is r_1 and from the other light is r_2:

$$\frac{P_1}{4\pi r_1^2} = \frac{P_2}{4\pi r_2^2}$$

and

$$\frac{P_1}{P_2} = \frac{r_1^2}{r_2^2}$$

The above equations let us calculate the distances at which two lights of different powers will balance. For example, if the second light is twice the power of the first, $P_2 = 2P_1$, then

$$\frac{P_1}{P_2} = \frac{1}{2} = \frac{r_1^2}{r_2^2}$$

and

$$\frac{r_1}{r_2} = \frac{\sqrt{1}}{\sqrt{2}} = \frac{1}{1.4}$$

The distance to the brighter light is 1.4 times the distance to the dimmer light.

Light is one of many phenomena that vary inversely with the square of the distance from the source. Other phenomena that follow an inverse square law include sound, magnetism, and gravity.

Credits

This snack is related to the Exploratorium exhibit Light Edge Photometer.

Palm Pipes

Make music that's entirely out of hand.

If you bang the open end of a piece of PVC pipe against the palm of your hand, you'll make a musical sound. The frequency, or pitch, of the sound depends on the length of the pipe. Based on this simple but significant fact, you can make instruments for your own pipe band. You'll find it surprisingly easy to play some simple songs.

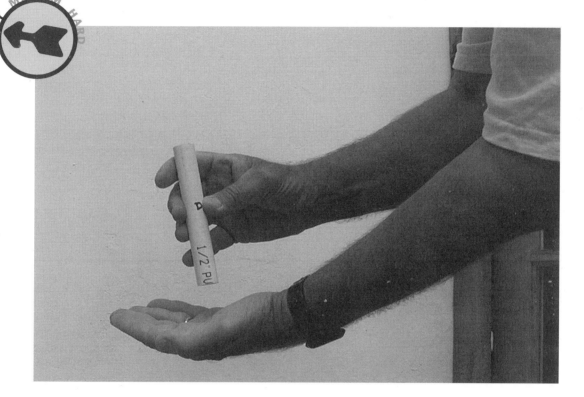

Materials

- PVC shears or hacksaw (Inexpensive PVC shears are amazingly handy for cutting PVC pipe and can be bought at hardware or home improvement stores; if available, an electric band saw, chop saw, or table saw makes the job of cutting many pieces of PVC go even faster.)
- $\frac{1}{2}$-in PVC pipe, about $6\frac{1}{2}$ ft (2 m) long
- centimeter ruler
- sandpaper (if necessary)
- permanent marker pen
- some friends

1 Cut a piece of PVC to each of the 15 lengths listed in the box shown here.

2 If the cutting process leaves sharp edges or fragments on the ends of any of the pipes, remove them with sandpaper.

Note	Length (cm)	Frequency (Hz)
F_1	23.6	349
G_1	21.0	392
A_1	18.7	440
Bb_1	17.5	446
C_1	15.8	523
D_1	14.0	587
E_1	12.5	659
F_2	11.8	698
G_2	10.5	784
A_2	9.4	880
Bb_2	9.2	892
C_2	7.9	1046
D_2	7.0	1174
E_2	6.2	1318
F_3	5.9	1397

3 Label each pipe with the musical note corresponding to its length. The subscripts in the list refer to the octave of the note. (The first note in the list is the F above middle C.)

To Do and Notice

Play the songs below by gathering a group of people, distributing one pipe to each person, and having each player sound his or her pipe at the right time. Before you begin, each person should practice making a good tone by holding the pipe vertically with one hand and banging the bottom end into the palm of the other hand. It's important not to cover the top end of the pipe.

Playing a song is easiest if there's one player per note, but if you have too few people, you can have one or more players take responsibility for two pipes. If you do this, however, make sure a person doesn't have to sound two different notes in a row. If you have many more people than there are notes in a song, you can have some players play pipes that are an octave lower than the ones called for. For example, when the F_2 pipe for "Twinkle, Twinkle, Little Star" is being played, another player can play the F_1 pipe at the same time.

You may find it helpful to have one person act as a conductor. A little practice will help a lot. Have fun!

Mary Had a Little Lamb
(Pipes C_1, D_1, E_1, and G_2: four players)

E D C D E E E D D D E G G
E D C D E E E E D D E D C

Twinkle, Twinkle, Little Star
(Pipes F_2 through D_2: six players)

F F C C D D C Bb Bb A A G G F
C C Bb Bb A A G C C Bb Bb A A G
F F C C D D C Bb Bb A A G G F

Jingle Bells
(Pipes C_1 through G_2: five players)

E E E E E E G C D E
F F F F F E E E E E D D E D G
E E E E E E G C D E
F F F F F E E E G G F D C

My Country 'Tis of Thee (America)
(Pipes E_1 through D_2: seven players)

F F G E F G A A Bb A G F G F E F
C C C C Bb A Bb Bb Bb Bb A G
A Bb A G F A Bb C D Bb A G F

What's Going On?

When you hit the open end of the pipe against your hand, air molecules in the bottom of the tube are squeezed together. This starts a process that takes four trips up and down the tube (see figure 1).

The molecules that have been squeezed together, in turn, squeeze the molecules next to them, and so on. In a sort of domino effect, the pulse of compression (high-pressure air) travels up the tube. When the pulse of compression reaches the top of the tube (see figure 1b), it expands outward into the air around the tube. In the process, some air molecules overshoot the end of the tube, producing a region of expansion (low-pressure air) in the top of the tube.

Air molecules just below the area of expansion rush upward to fill it, creating a pulse of expansion that travels back down the tube. When this pulse reaches the bottom, it reflects off your palm and travels back up the tube as another pulse of expansion (see figure 1c).

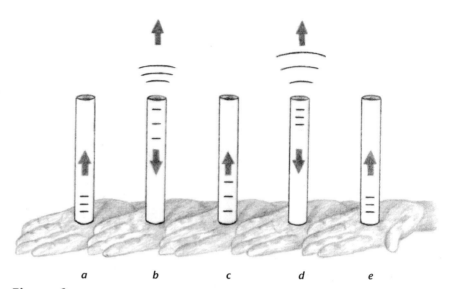

Figure 1 *Slap the pipe with your hand: (a) a compression pulse travels up; (b) the compression reflects from the open end as an expansion; (c) the expansion reflects from your hand; (d) the expansion reflects from the open end as a compression; and (e) the process repeats.*

When it reaches the top, some air from outside the tube rushes into the low-pressure area, creating an area of compression, which travels as another pulse back down the tube (see figure 1d).

When this pulse of compression reaches the palm of your hand (see figure 1e), it reflects, and at this point the whole process repeats itself.

A pulse that starts at your palm as a compression makes four complete transits of the tube (up as a compression, down as an expansion, up as an expansion, and down as a compression) before one whole cycle is completed. This four-part cycle corresponds to one wavelength of a sound, or one single vibration. A series of these repeated cycles is the source of the sound you hear when you "play" one of the pipes.

The length of the tube affects the note that the tube produces. Because the speed of sound waves is the same in all the tubes, the length of the tube has a direct effect on the time it takes for a compression-expansion pulse to make its four transits of the tube. The longer it takes for a pulse to complete its cycle and start over again, the fewer the cycles, or vibrations, per second. The fewer the vibrations per second, the lower the frequency of the sound, and the lower the musical note. Thus, long tubes produce lower notes, and short tubes produce higher notes.

So What?

Instruments with a long tube, such as a bass saxophone, produce lower frequency notes than instruments with shorter tubes. Although modern symphony instruments may seem quite complex, the basic relationship between long and short tubes is the same as exists for the simple palm pipes in this snack.

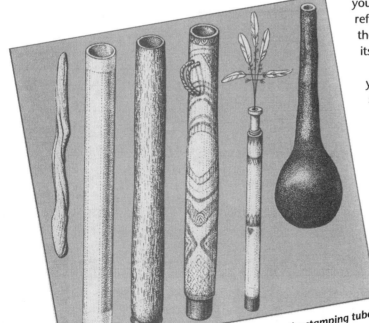

The palm pipes used in this snack are related to the stamping tubes, stamping stick, and gourd shown in this picture. These, in turn, are part of a larger family of instruments known as idiophones, which includes rattles, bells, gongs, and xylophones, among others.

Twinkle, Twinkle, Little Star
(Pipes F_2 through E_2: three players on Melody [M], and three players on Harmony [H])

M:	F	F	C	C	D	D	C	B♭	B♭	A	A	G	G	F
H:	C	C	A	A	B♭	B♭	A	G	G	F	F	E	E	C
M:	C	C	B♭	B♭	A	A	G	C	C	B♭	B♭	A	A	G
H:	A	A	G	G	F	F	C	A	A	G	G	F	F	C
M:	F	F	C	C	D	D	C	B♭	B♭	A	A	G	G	F
H:	C	C	A	A	B♭	B♭	A	G	G	F	F	E	E	C

Going Further

Two Voices

Have one group of players play the melody notes (M) of "Twinkle, Twinkle, Little Star," while another group plays the harmony notes (H) shown below each melody note.

If It Hertz

You can calculate the approximate frequency of sound that any length of pipe will produce. Here's how:

The velocity of a sound wave (v) is equal to its frequency (f) times its wavelength (λ). Rearranging this equation gives $f = v/\lambda$. The value for v is about 350 m/s, the speed of sound in air around room temperature. The value for λ can be obtained by multiplying the tube length (in meters) by 4 (which is the number of transits a compression-expansion pulse makes inside the tube for one sound wave). Therefore, if you divide 350 m/s by a tube's λ value, you obtain the approximate frequency (in cycles per second, or *hertz*) of the note the tube will produce. (Exact values involve additional considerations, such as the diameter of the tube.)

The Difference a Diameter Makes

Reflection of a sound wave doesn't occur exactly at the open end of a tube but instead happens at a point slightly beyond the end. The larger the diameter of the pipe, the farther from the end the reflection occurs. To more accurately estimate the value for the frequency, add 0.3 of the inside diameter to the length of the tube. Try this snack with pipes of different diameters.

Credits & References

This activity was developed by Gene Easter and Bill Reitz. Art Fortgang first brought it to our attention.

Hopkin, Bart. *Musical Instrument Design.* Tucson, Ariz.: See Sharp Press, 1996. An invaluable reference for designing and experimenting with homemade musical instruments.

Periscope with a Twist

Around and around and around it goes.

Periscopes are standard equipment on submarines, but a submarine captain might go crazy using one like this. It is constructed so that the mirror on one end can be rotated. If you look into the bottom mirror and then rotate the top mirror, you can see off to the side and in back of you, but things will also turn sideways and upside down.

Materials

- template
- cardstock, 2 pieces of $8\frac{1}{2}$ in × 11 in (available at copy shops or office supply stores; use the heaviest cardstock that can be run through the copier you are using)
- ballpoint pen
- ruler or straightedge
- scissors
- utility knife or X-Acto knife

- transparent tape
- double-stick tape
- plexi-mirror, 2 pieces, $1\frac{3}{4}$ in × $2\frac{1}{2}$ in (5.1 cm × 6.4 cm). Plexi-mirror is plastic mirror, which is available at plastics stores (look in the yellow pages under Plastics for locations); plexi-mirror can be cut with a band saw, or with a plywood blade on a table saw, or the plastics store can cut it for you.

See Alternative Construction for use of glass mirror instead of plexi-mirror.

- $\frac{1}{2}$-in PVC pipe, Schedule 125 or Schedule 40, 1 piece, about 12 in (30 cm) long. Schedule 125 PVC is thinner walled than the more common Schedule 40 and gives a little larger view, but Schedule 40 can also be used. See Alternative Construction for use of file folders instead of PVC pipe.

1 Photocopy the end-cap template (see figure 13 on the last page of this snack) onto cardstock. Make two copies.

2 Use the ballpoint pen and straightedge to trace over all the lines on both templates that are marked "Fold up" or "Fold down." Press hard enough to make a slight indentation in the material. This indentation will make it easier to fold the material.

3 Cut an entire template out in one piece along its outside edges.

4 Cut along all the lines marked "Cut." This will divide the template into two pieces: the main section and a separate strip with parts F and G.

5 Fold the B and C flaps up as shown in figure 1.

Figure 1

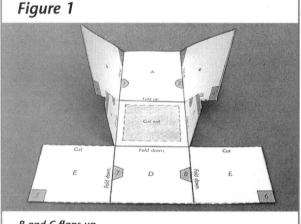

B and C flaps up

6 Fold the strip with parts D and E down as shown in figure 2.

Figure 2

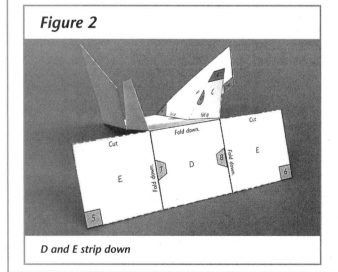

D and E strip down

7 Fold each of the two parts E down (actually, back) as shown in figure 3.

Figure 3

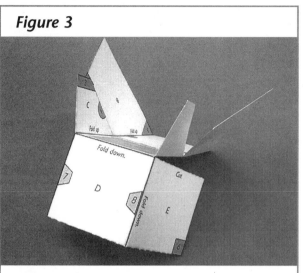

E flaps down and back

8 Fold part A up as shown in figure 4.

Figure 4

Flap A up

9 Once all the folds have been made, cut out the shaded section labeled "Cut out." (It may be best to lay the template flat again and use a utility knife or X-Acto knife, but scissors can also be used.)

10 After the "Cut out" section has been removed, re-fold the structure as shown in figure 5. The numbered shaded areas on parts A, B, and C should register with each other as closely as possible, as shown in figure 6. Use transparent tape to tape part C to part B on both sides.

Figure 5

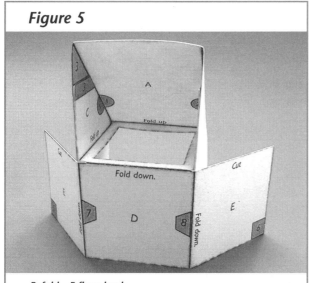

Refold—E flaps back

Figure 6

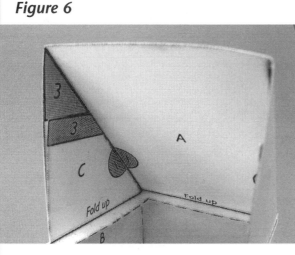

Close up of registration marks

11 Fold part E back on each side so that it is up against the outside of part B, with the two shaded squares numbered 5 back-to-back on one side, and the squares numbered 6 back-to-back on the other side, as shown in figure 7. Use transparent tape to tape part E to part B on both sides.

Figure 7

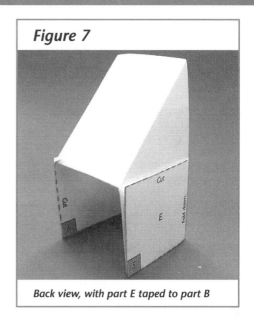

Back view, with part E taped to part B

12 For the separate piece with parts F and G, fold the G parts to make a U-shaped piece, and place it so that part F will be in a position to cover the opening in the back of the cap structure as shown in figure 8.

Figure 8

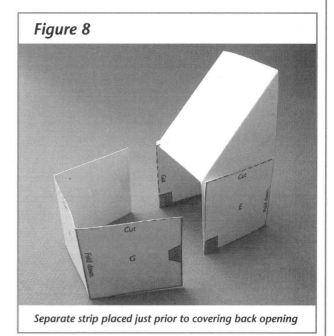

Separate strip placed just prior to covering back opening

13 Place the U-shaped piece so that it covers the back opening and the two shaded areas numbered 7 and the two shaded areas numbered 8 register with each other on the front as closely as possible, as shown in figure 9. Use transparent tape to tape part G to part D on both sides.

14 Use additional transparent tape as necessary to securely hold the end cap together.

Figure 9

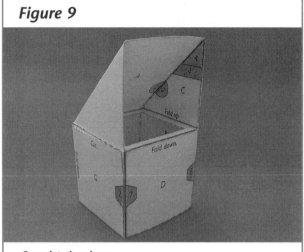

Fold up

Cut

Fold down

Completed end cap

15 Repeat steps 3 through 14 to make a second end cap.

16 Place double-stick tape on the back of each mirror, and attach one mirror inside each end cap by pressing it onto the inside surface of part A.

Alternative Construction

- Instead of using PVC pipe for the tube, you can cut a file folder in half along the fold, and use one half to make a tube that has a diameter of $1\frac{7}{8}$ in (4.8 cm). This is the same as the outer diameter of $1\frac{1}{2}$ in PVC pipe, the size for which the end cap template has been designed.

- Glass mirror may be used instead of plexi-mirror, but it has sharp edges, is subject to breakage, and may not hold as well to the tape. One-foot-square glass mirror tile is commonly available at home improvement stores and can be cut with a glass cutter. **CAUTION:** Be sure to take proper safety precautions if you do this, and get advice from someone with experience if you have never tried using a glass cutter.

17 Place one of the end caps on one end of the PVC pipe and tape it in place. This will be the bottom mirror. Place the other end cap on the other end of the PVC pipe without taping it so that it is free to rotate. This will be the top mirror.

Using periscopes for a better view, spectators watch the 1962 National Open.

To Do and Notice

Turn the top mirror so that it is facing forward in relation to the bottom mirror. Hold the periscope up to your eye and look through the bottom mirror. Observe your surroundings through the periscope, keeping the top mirror facing forward.

While holding the periscope up to your eye with one hand, use the other hand to slowly rotate the top mirror housing until you have looked in a complete circle. Don't rotate your head or body; keep your head and the bottom mirror always pointing in the same direction.

Notice that when the mirror faces sideways, what you are looking at is lying on its side, and when you are looking backwards, what you are looking at is upside down. As you rotate the mirror, the world around you rotates in a full circle.

Have someone stand in front of you. Look at that person's head through the periscope with the top mirror facing forward. Ask the person to move slowly in a circle around you and to remain facing you the whole time. Slowly rotate the top mirror to follow the person's movement. When the person is off to your side, his or her head will be horizontal, and when the person is in back of you, his or her head will be upside down! As you rotate the mirror, the person's head rotates in a full vertical circle. (The person being looked at can see your head rotate as well.)

Try looking at other things around you while rotating the periscope.

Again, hold the periscope up to your eye and look through the bottom mirror, with the top mirror facing forward. But this time, rotate the whole periscope (and your head and body with it) in a full circle while looking into the bottom mirror. Notice that this time there is no rotation of what you are looking at. Everything is always right-side up.

What's Going On?

When the top mirror is facing forward, the image is rotated 90 degrees by the top mirror, and then is rotated back 90 degrees in the opposite direction by the bottom mirror, as shown in figure 10. What you see is an image in the same orientation as the original. When the top mirror is facing backwards, however, the image is rotated 90 degrees by the top mirror, and then another 90 degrees in the same direction by the bottom mirror, as shown in figure 11. What you see then is an image that is flipped 180 degrees from the original. Mirror positions in between these two extremes result in an intermediate amount of rotation of the image. For example, when the top mirror is facing sideways, as shown in figure 12, the image is horizontal.

So What?

You may have seen movies or videos of submarine commanders using periscopes. If you think about it, the person using the periscope usually walks around with the periscope, rather than standing still. If only the top mirror on the periscope moved, then only things that were straight ahead would be right side up. Ships seen through the periscope would sometimes be sideways or upside down!

Did You Know?

Wartime Periscopes

World War II–vintage submarine periscopes were not as simple as this one. They used a complex series of lenses and prisms. Simpler reflecting periscopes, however, were used in tanks. Prisms were used instead of mirrors, and the periscopes were fixed in position, serving as a kind of window. Some of these tank prisms can still be purchased today from companies that sell military surplus items.

Going Further

The Other Side of the Fence

Try making a periscope with a long tube that you can use to see over a high fence. You can improvise a longer tube from the cardboard core of a wrapping-paper roll. If you were able to make a periscope with a very long tube, what would happen to the amount of area that you were able to view? (Hint: Try looking at a particular area through a toilet-paper tube and then through a longer paper-towel tube of the same diameter.)

Credits

Don Rathjen, Barbara Ziegenhals, and Glorianne Hirata were involved in the development and evolution of this snack.

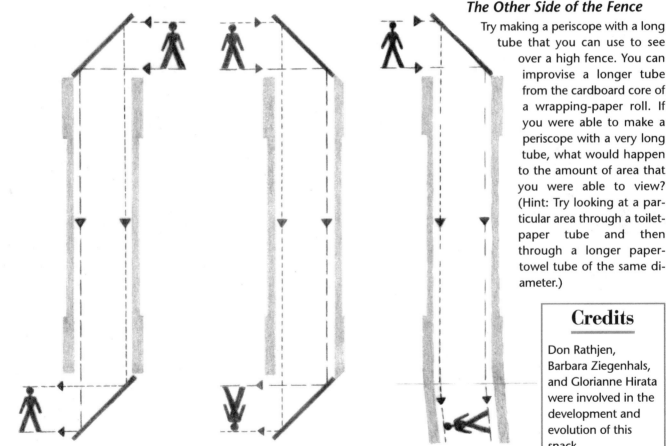

Figure 10 *Rotation of the image when the top mirror faces forward.*

Figure 11 *Rotation of the image when the top mirror faces backward.*

Figure 12 *Rotation of the image when the top mirror is facing sideways.*
A diagram for an intermediate position such as this is harder to understand because it must show rotations that occur in three dimensions.

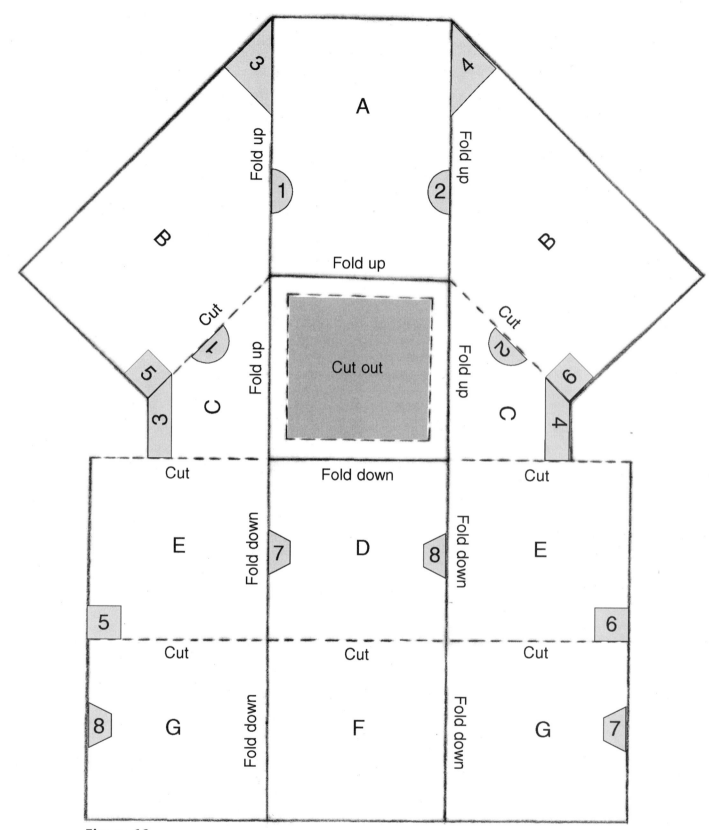

Figure 13 **End-cap template**

Personal Pinhole Theater

A pinhole camera you can really get into.

Have you ever heard of a camera without a lens? Or without film? In this snack you put your head in a large, dark box with a hole poked in the side. You view an image on the inside wall of the box that is essentially the same as the image formed on the film in a regular camera, except that there will be no permanent record of it.

Materials

- masking tape
- white paper, a few sheets
- large cardboard box (large enough to give you a foot of headroom inside when you rest it on your shoulders)
- utility knife
- scissors
- aluminum foil
- duct tape
- pushpin
- dark sweatshirt or towel
- pencil

1 Before you read on, take a good look at figure 1 to get an idea of what you will be doing with the box.

2 Tape sheets of white paper onto the inside wall of the box that you will be facing when your head is in the box. This will be the viewing screen.

3 Cut a small opening approximately 3 inches square (7.5 cm × 7.5 cm) in the side of the box opposite the screen, well above where your head will be (see figure 1). This is where the pinhole will be located.

4 Use the knife to cut a hole in the bottom of the box, through which you can put your head. Position this hole so that the back of your head will rest against the wall that is opposite the screen. The hole should be just big enough for your head to slip through, as shown in figure 1.

5 Cut a flat square of aluminum foil about 4 inches × 4 inches (10 cm × 10 cm). Make sure this piece of aluminum is large enough to cover the opening you cut in step 3. Use masking tape to tape the piece over the opening.

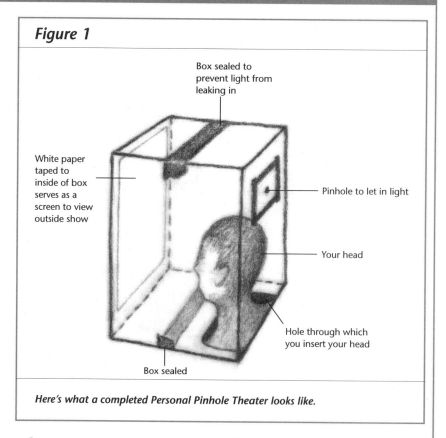

Figure 1

Box sealed to prevent light from leaking in

White paper taped to inside of box serves as a screen to view outside show

Pinhole to let in light

Your head

Hole through which you insert your head

Box sealed

Here's what a completed Personal Pinhole Theater looks like.

6 Put the box on your head and check the corners and seams for light leaks. Use duct tape (or masking tape and aluminum foil) to seal any leaks you find. Don't worry for now if light is leaking in around your neck.

7 Use the pushpin to poke a pinhole in the center of the piece of aluminum foil. (A pushpin is best as it has a larger diameter than a straight pin.)

To Do and Notice

Find a well-lit area (outside on a bright day is great) and put your head into the box. Position your head so that you are facing the screen. Wrap the sweatshirt or towel around your neck to keep light from leaking into the box from the bottom. As your eyes adjust to the dim surroundings, notice if anything appears on the screen.

When images do appear, notice their orientation. Are they right-side up or upside down? Are they left-right reversed or normal?

Move around and notice how the images on the screen change. Try to position yourself so that two similar objects at different distances away (for example, two cars) cast their images on the screen at the same time. Which image is larger? Are both images in focus at the same time?

After you have had a chance to use your Personal Pinhole Theater for a while, try enlarging the pinhole significantly by carefully poking a pencil through it until it is about half to two-thirds the diameter of the pencil. What happens to the image on the screen as a result? (If you want to go back to the small hole, you can tape foil over the large hole and poke a new small hole with the pushpin.)

What's Going On?

You have made a *camera obscura* ("dark room"), or pinhole camera. Any images on the screen are upside down and left-right reversed. The drawing in figure 2 shows how the up-down part of this reversal takes place.

Light rays are coming from every point on the tree and hitting the outside of the box (light rays from only

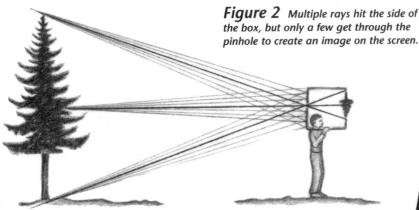

Figure 2 *Multiple rays hit the side of the box, but only a few get through the pinhole to create an image on the screen.*

At the Exploratorium, a thin slit between two doors creates pinhole images of the columns outside the door. The images are on the floor of the museum.

three points are shown on the drawing because it would be impossible to show all of them). Each light ray in effect carries an image of the point on the tree where it originated, and so the entire side of the box is covered with images of all points on the tree. The result is an unintelligible overlapping of images.

The pinhole, however, lets only a small number of rays from each point pass through, and the rays from each point are projected on a small area of the white screen, without significant overlap from the images of the other points. The result is a clear image of the tree. As you can see, however, the ray from the top of the tree hits the lower portion of the screen, and the ray from the bottom of the tree hits the upper portion. This geometry results in the image of the tree appearing upside down on the screen (and left-right reversed, as well).

The farther an object is from the pinhole, the smaller its image will be on the screen, as shown in figure 3. Focus (sharpness of the image) is not affected by distance.

Using a larger pinhole gives you a brighter image, since it lets in more light rays. But this increases the overlapping of images, causing the image to lose sharpness and become blurry.

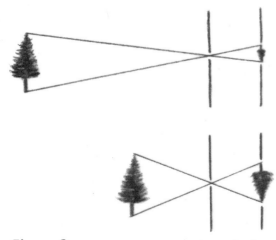

Figure 3 *This ray-tracing diagram shows why the image on the screen is bigger when an object is closer to the pinhole.*

So What?

The Personal Pinhole Theater demonstrates a classic principle of optics, one that photographers must take into account in every shot: There's a trade-off between brightness and sharpness.

In photography, the zone in which objects will be in sharp focus is known as *depth of field.* The smaller the lens aperture, or opening, of a camera, the greater the depth of field. In other words, the more the lens opening is like a pinhole, the greater the ability to have objects at different distances from the camera in focus simultaneously. As the opening is made smaller, however, less light is allowed through, and you have to use a longer exposure time to compensate. To take a picture in dim light, you open the lens wider to let in more light, but in so doing you sacrifice depth of field.

Going Further

Pinhole Variations

Systematically vary the size or shape of the pinhole you use and note how the image changes. You might also experiment with using two pinholes.

Walk Right In!

Make another Personal Pinhole Theater out of a refrigerator box so that it is big enough for your whole body.

Real Photographs

You can build a pinhole camera that will actually capture photographic images on film. One source for this activity is Jim Shull's book in the Credits & References section, and others can be readily located.

Did You Know?

Artist's Tool

Renaissance painters used portable camera obscuras very much like your Personal Pinhole Theater, except that they were smaller and handheld, and the screen side had only a piece of translucent paper, rather than a solid opaque wall. The image projected on the translucent paper could be seen from outside the box and could be traced directly onto the paper. (Sometimes a black cloth was draped over the screen and the user's head, to prevent the image from getting washed out by normal daylight.) The traced drawings of objects helped the artist represent perspective realistically. Johannes Vermeer (1632–1675) is one artist of note who is thought to have used a camera obscura in his work (see figure 4).

Popularly known as the Giant Camera, a room-sized camera obscura can be found near the Cliff House in San Francisco (see figure 5). Double convex lenses in the ceiling of the camera-shaped building rotate 360 degrees every six minutes, reflecting external images of the surrounding cliffs and Pacific Ocean onto a large concave bowl wherein visitors can easily discern seagulls in the sky, seals on the rocks, and tourists walking by outside.

Figure 5 Near San Francisco's Cliff House restaurant at Land's End is a room-sized camera obscura open to the public on clear days.

Figure 4 View of Delft, *by the Dutch painter Johannes Vermeer, may have been created using a camera obscura to trace the outlines of the scene, enabling Vermeer to create a painting that is almost photographically precise.*

Credits & References

This snack was developed by Eric Muller of the Exploratorium Teacher Institute. It's related to the Exploratorium exhibits Holes in a Wall and Cracks in a Door.

Chevalier, Tracy. *Girl with a Pearl Earring.* New York: Plume, 2001. A novel centered around Vermeer's painting "Girl with a Pearl Earring," which mentions Vermeer's use of a camera obscura loaned to him by his friend, Antonie van Leeuwenhoek, an early developer of the microscope.

Shull, Jim. *The Beginner's Guide to Pinhole Photography.* Buffalo, N.Y.: Amherst Media, Inc., 1999.

Steadman, Philip. *Vermeer's Camera: Uncovering the Truth Behind the Masterpieces.* Oxford: Oxford University Press, 2001.

Also see the Pinhole Magnifier snack in the *Exploratorium Science Snackbook,* 1991.

Perspective Window

You'll be drawn to it.

Renaissance artists used a device like this to learn how to draw objects in perspective. You'll use the Perspective Window in combination with the mathematics of similar triangles to find the height of an object.

Materials

- scissors (if the cardboard is too thick to cut with scissors, you may have to use a utility knife)
- stiff cardboard (regular poster board is not quite stiff enough) or foam core, approximately 6 in × 12 in (15 cm × 30 cm)
- ruler or tape measure

- double-stick tape, $\frac{1}{2}$-in-wide (12-mm) roll
- CD jewel case
- 2 rubber bands
- chenille stem (shorter pipe cleaners can also be used, but you'll have to twist two of them together)

- masking tape
- overhead projector pen (not permanent)
- damp paper towel or tissue
- dry paper towel or tissue
- pencil
- a friend

Figure 1

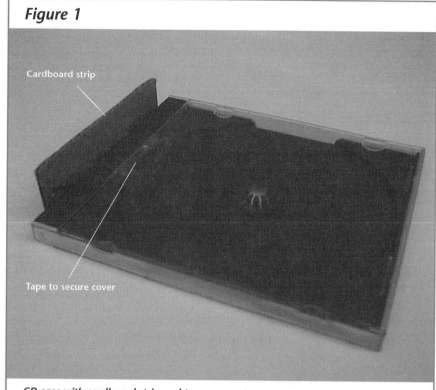

Cardboard strip

Tape to secure cover

CD case with cardboard strip and tape

two rubber bands to hold the cardboard in place as shown in figure 2.

6 Place one of your fingers about 4 inches (10 cm) from the end of the chenille stem. Wrap the stem around your finger to make a loop, and twist the rest of the end around the stem to create a circular "eyepiece."

7 Bend the pipe cleaner 90 degrees into an L shape, so that the center of the eyepiece is about 3 inches (7 cm) from the bend, and then make another 90-degree bend about an inch from the straight end. Twist the stem as necessary, and tape it to the base with masking tape as shown in figure 2. Be sure that the eyepiece is parallel to the face of the raised cover, which serves as the "window."

1 Cut one piece of cardboard 4 inches × 11 inches (10 cm × 28 cm), and another piece 4 inches × 1 inch (10 cm × 2.5 cm).

2 Use a 4-inch (10-cm) piece of double-stick tape to attach the small cardboard strip to the end of the CD case near the cover hinge, so that it will prevent the clear cover from opening more than 90 degrees (see figure 1).

3 Put a piece of double-stick tape about 2 inches (5 cm) long on the cover, along the edge that contacts the vertical piece of cardboard (see figure 1). When the lid is raised, this piece of tape will stick to the cardboard and keep the lid from falling closed.

4 Open the cover of the CD case so that it is vertical, and press the small piece of tape against the cardboard to hold the cover in place.

5 Place the larger piece of cardboard on the inside surface of the base of the CD case so that one end is up against the open cover. (There will be a small space between the edge of the cover and the base; it is okay if the cardboard slips through this space to rest against the other piece of cardboard.) Use the

Figure 2

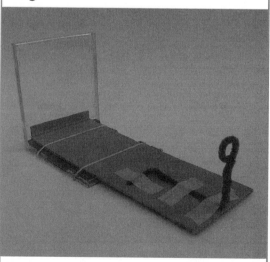

Complete assembly

To Do and Notice

Set your Perspective Window on a level surface so that you can comfortably keep your eye near the eyepiece. (You can hold it in your hand instead, but it may be hard to hold steady, and your arms will get tired.)

Part 1: Make a Scale Drawing of an Object

Look through the eyepiece and window at an object that fills up most of the window (the object can be large and in the distance, or small and fairly close). Hold the device steady with one hand; with your other hand, use the overhead projector pen to draw on the window the outlines and features of whatever you are looking at. What you draw on the window is a scaled-down drawing of the object you have chosen.

If you want to do a different drawing, wipe the window with the damp paper towel or tissue to remove the previous drawing, and then dry it.

Part 2: Measure a Friend's Height Indirectly

Have a friend stand still in your line of sight so that his or her total height can be seen through the window. Carefully draw your friend's image on the window. Before your friend moves,

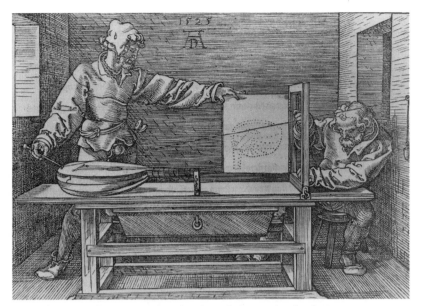

A Renaissance artist uses a perspective view window to sketch a lute.

measure D, the distance from your eye to his or her feet. Then measure d, the distance from your eye to the window, and h, the height of the image you drew on the window.

Using a proportion created by the similar triangles illustrated in figure 3, calculate your friend's height:

$$\frac{h}{d} = \frac{H}{D}$$

Check your calculated value for H by measuring the height of your friend directly.

> ⮕ **Helpful Hint**
>
> When you are done using the Perspective Window and close the CD lid, make sure the tape on the cover comes unstuck from the cardboard and doesn't pull the cardboard off the case. If you open and close the cover many times, you may need to replace the piece of tape on the cover.

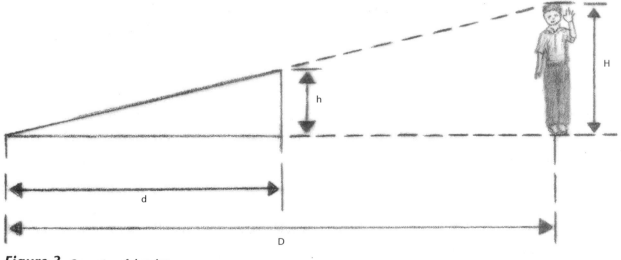

Figure 3 *Geometry of drawing*

What's Going On?

There's an imaginary triangle between you and any object in your field of view. One of the vertices of the triangle is your eye, and the other two are any two points on the object in a vertical plane, such as its top and bottom.

When you trace an object as you look through the eyepiece of the Perspective Window, you're creating another triangle, smaller than and similar to the larger one. The two triangles have the same angles and proportional sides.

Therefore, the height-to-distance ratio of any object you view will be proportional to the ratio of the height of your drawing and the distance from the window to your eye.

So What?

It may seem cumbersome to use the Perspective Window to measure the height of your friend indirectly when it is easier to do it directly. However, this device can also be used to find the height of something you *can't* measure directly, such as a tree, a building, or a flagpole. If you can find h (the height of the drawing of the object), d (the distance from the window to your eye), and D (the distance from your eye to the object), you can find H, the height of the object, using the proportion in part 2 of To Do and Notice.

You can also use the Perspective Window to find the distance between you and an object if you know the object's height. As long as you know three of the four values in the proportion, you can solve for the unknown value.

Indirect measurements allow us to estimate the size or distance of objects that we cannot measure directly. Many measurements made by astronomers, for example, must be made indirectly.

Going Further

Perspective Challenge

Draw three squares in a row on the ground with some chalk, each about 24 inches (60 cm) on a side, or draw smaller squares on a piece of paper. Place your Perspective Window so that it's on the line formed by the squares and you can see all three squares through the eyepiece. Draw the three squares on the window. The drawing on your window won't look like three squares—instead, it will look like three diminishing trapezoids. What would you need to draw on the ground (or paper) so that the picture on your window looks like three squares?

Playing with Proportions

What do you think will happen to the size of a scale drawing of an object if you make the drawing with the eyepiece closer to the CD cover window? Try it: Make a scale drawing of an object. Then (without moving the CD case) remove the eyepiece from the cardboard base and remove the base itself from the CD case. Retape the eyepiece to the inside bottom of the CD case, so that the vertical part of the eyepiece is even with the edge opposite the window.

Measure the new distance from the eyepiece to the window; it should be $5\frac{1}{2}$ inches (14 cm), or half the original distance. Now make another scale drawing of the same object. Measure the heights of the two drawings. Can you see a relationship between the heights of the two drawings and the two distances from eyepiece to window?

Credits

This snack is based on the Exploratorium exhibit of the same name. Lori Lambertson contributed to this adaptation.

Pinhole Mirror

Image may not be everything, but for the sun it tells you a lot.

You can't look at the sun directly, but with this snack you can shine an image of the sun onto a viewing screen. If you keep the mirror absolutely still, you will be able to use the reflected image to track the sun's motion and find the angular diameter of our local star.

Materials

- hot glue gun and glue sticks
- 2 pieces of $\frac{1}{2}$-in PVC pipe, each approximately 6 in (15 cm) long
- piece of wood, approximately $\frac{3}{4}$ in × 4 in × 8 in (2 cm × 10 cm × 20 cm), exact size not crucial (you can use standard 1 in × 4 in shelving, which is actually $\frac{3}{4}$ in thick and $3\frac{1}{2}$ in wide)
- 2-liter plastic bottle with cap
- water
- double-stick tape

- small piece of mirror of any shape, glass or plexi-mirror (Plexi-mirror is plastic mirror and is available at plastics stores, which you can locate through the yellow pages of your phone book; it can be cut with a band saw, or with a plywood blade on a table saw, or you can get it cut at the store.)
- masking tape
- sunlight

- foam core or poster board about 24 in × 24 in (60 cm × 60 cm), exact size not crucial
- stopwatch, or watch with second hand
- pencil and paper

Alternative Construction

You can skip Assembly step 1 and use two large books as a stand instead of making one.

1 Make a stand for the bottle: Hot-glue the PVC pipes to the wood base, parallel to the long edges of the base, about $1\frac{1}{2}$ inches (4 cm) apart (see figure 1). Use a generous bead of glue so that the pipes will stick well.

2 Fill the bottle with water, and put the cap on tight.

3 Cover all but a $\frac{1}{4}$ inch $\times \frac{1}{4}$ inch (0.5 cm $\times$ 0.5 cm) square in the cen-

Figure 1

Bottle with mirror attached and stand

ter of the mirror with masking-tape strips, as shown in figure 2. Use a small piece of the double-stick tape to attach the mirror, shiny side out, to the outside of the bottle, approximately centered on the bottle's length. (The mass of the water will anchor the bottle firmly in place and help keep the mirror from vibrating.) Figure 3 shows what the completed assembly should look like.

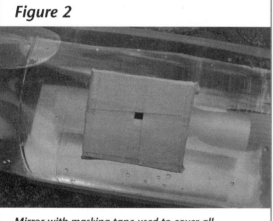

Figure 2

Mirror with masking tape used to cover all but a small square

Figure 3

Bottle and mirror on stand

To Do and Notice

On a sunny day, take the bottle, the stand, and the piece of poster board or foam core to use as a viewing screen outside, along with a watch, a pencil, and some paper to write on.

Place the bottle on the stand so that the mirror faces the sun directly. Hold the screen about 3 feet (1 m) away from the mirror, oriented so that its flat side faces the mirror, but not so that it prevents the sun from hitting the mirror. Rotate the bottle, move the base, and adjust the position of the

screen until you get an image of the sun to appear on the screen. **CAUTION: Never look directly at the sun!**

You'll get the best results when the mirror is approximately perpendicular to the sun's rays, and the viewing screen is approximately parallel to the mirror and off to the side a little from the line between the sun and the mirror (see opening photo).

Observe the image of the sun on the screen.

Find a place where you can lean the screen against some fixed object, such as a tree or a wall and where you can position the mirror to project the

sun's image on the screen. Ideally, the screen should be in the shade, so that the sun's image has the most contrast with the screen. While holding the screen firmly in place so it doesn't move, quickly trace the image of the sun on the screen, and then immediately start the stopwatch (or note the time to the second). Stop timing when the image of the sun has moved completely out of the circle you have traced. Record the length of time that elapsed. Repeat the process of tracing and timing several times.

Increase the size of the image by moving the screen several feet (a meter or two) farther from the mirror. You may have to raise the screen off the ground or adjust the position of the bottle and stand if the image gets too high in the air or is otherwise out of reach or blocked. Repeat the process of tracing the image and timing how long it takes the image to move completely out of the circle.

What's Going On?

If you held a sheet of paper with a pinhole in it between the sun and a white screen, an image of the sun would appear on the screen as shown in figure 4. Notice that light from the "top" of the sun passes through the pinhole to form the *bottom* part of the image, and that light from the "bottom" of the sun passes through to form the *top* of the image. The image of the sun on the screen is inverted, and the rays of light from sun to pinhole and from pinhole to image form two similar triangles.

In this snack, the mirror acts just like a pinhole. It reflects an inverted image of the sun onto the viewing screen, as shown in figure 5. In addition, as with a pinhole, the rays of light from sun to mirror and from mirror to image form two similar triangles.

The earth's rotation causes the image of the sun to appear to move across the screen. What is the angular velocity of the earth's spin? If the earth spins 360 degrees per 24 hours, then it spins 15 degrees per hour, or 0.25 degrees per minute.

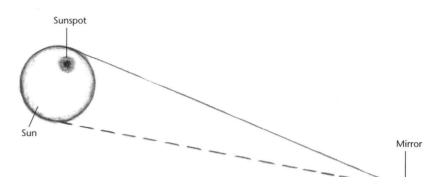

Figure 5 *Image of sun formed by light reflecting from a very small mirror*

Your timings should tell you that it takes about 2 minutes for the sun's image to move "one sun diameter" on your screen—whether the image is small or large. The time is constant because it's a measure of another constant, the rotation of the earth.

If the earth spins 15 degrees every 60 minutes, and in 2 minutes this spinning is equal to the sun's diameter, then we can find the diameter of the sun in degrees (its *angular* diameter) using this proportion:

$$\frac{15 \text{ degrees}}{60 \text{ minutes}} = \frac{x \text{ degrees}}{2 \text{ minutes}}$$

$$60x = 30$$

$$x = 0.5$$

The angular diameter of the sun is 0.5 degrees.

The angular diameter tells us how many degrees of angle the sun occupies in the sky. However, it doesn't tell us whether the sun is a small object close to us or a large object far away from us.

So What?

On a sunny day, look at the pattern of light and shadow on the ground under a large leafy tree. You may notice that many of the bright spots are small round circles or ellipses. These are images of the sun! Despite the irregular nature of the gaps in the leaves through which sunlight is able to pass, the light forms circular patterns on the ground. Each gap in the leaves acts as a pinhole, and as long as the gaps are reasonably small, the images are reasonably round. If the gaps are too large, then many overlapping images form, and the overall shape may no longer be circular. See Bigger Than a Pinhole, and Size or Shape? in the

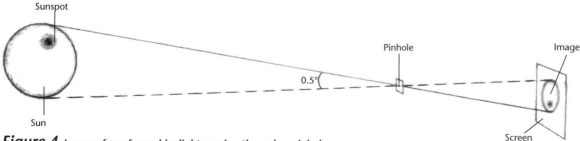

Figure 4 *Image of sun formed by light passing through a pinhole*

Going Further section for some things you can do to extend this idea.

Going Further

Solar Observatory

Try this snack at different times of the day and different times of the year. You can also use it to safely observe eclipses and sunspots.

Bigger Than a Pinhole

Try this snack using a mirror with a much larger unmasked square, approximately 2 inches × 2 inches (5 cm × 5 cm), in the middle. Can you get a round image of the sun on your screen? Why or why not?

Size or Shape?

Why is it possible to get a round image of the sun using a square reflecting surface? Try varying the shape (rectangle, triangle, irregular shape) of the unmasked part of the mirror while keeping the size very small. See if the shape of the mirror makes a difference in the image cast.

This multiple-exposure image shows the motion of the sun around the sky at the earth's South Pole. The sun circles completely around the sky staying a constant height above the horizon.

Credits

This snack was developed by Lori Lambertson of the Exploratorium Teacher Institute.

Reverse Masks

Eye'll be seeing you.

One mask protrudes from the black surface like an ordinary face, and the other is indented into the surface. When you close one eye and view the two masks, they both look like they are protruding, and when you move sideways, the indented mask seems to turn to follow your movement!

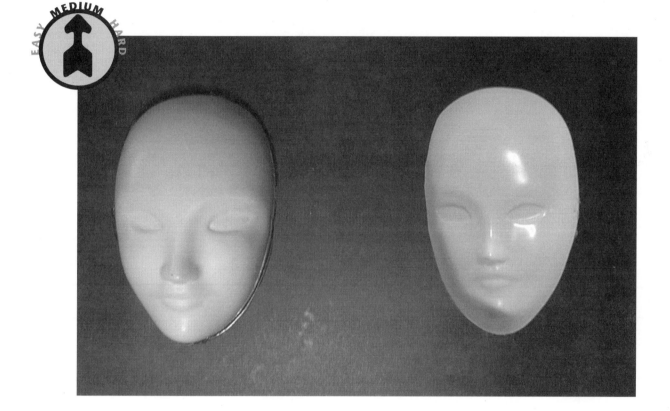

Materials

- 2 identical masks (Craft stores have relatively inexpensive blank masks that are white on both sides; Halloween masks also work well. If you can't get two identical masks, you can use two different masks.)

- masking tape (only necessary if you want to cover the eye and mouth holes of the masks)
- black poster board
- ruler
- pencil
- utility knife or sharp scissors

- hot glue gun and glue sticks
- movable light source (e.g., adjustable desk lamp)

If the masks you choose are not already white on both sides, you may need these materials as well:

- white spray paint, or white paint and a brush

1 You may first need to make some decisions about whether or not to modify your masks. If the masks aren't white on both sides, you may want to paint them so that both sides are white. If the masks have holes for the eyes, nostrils, and mouth, you may want to cover the holes with masking tape. (Put the tape on the "back" side of the mask, relative to the side you will be looking at.) Masks with holes and masks with colored outside surfaces will both work, but they may distract from the main effect of the snack.

2 Cut a sheet of black poster board on which you can place the masks side by side, with at least $2\frac{1}{2}$ inches (6 cm) between the masks, and a border of at least $1\frac{1}{2}$ inches (4 cm) around the edge.

3 Draw a vertical line dividing the poster board in half.

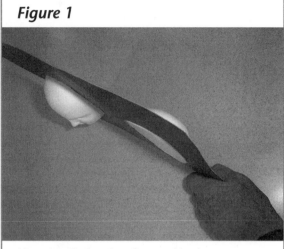

Figure 1

Mounting tilted to show front and back

4 Place one of the masks in the middle of the left half of the poster board, and carefully draw its outline.

5 Using the scissors or utility knife, carefully cut along the mask outline to make a hole in the poster board that is exactly the shape of the outline of the mask.

6 Place the mask right-side up on a table. Place the poster board over the mask so that the hole in the poster board fits over the mask. Press down on the poster board until it is flat on the table, with the mask protruding upward. If you have drawn and cut accurately, the edge of the hole should fit fairly closely around the edge of the mask.

7 Run a small bead of hot glue completely around the edge of the mask to glue it to the poster board.

8 Turn the poster board over, so that the mask is now on the right side and is concave (or "inside out").

9 Lay the second mask in the middle of the left side of the poster board, and run a small bead of hot glue completely around the edge of the mask to glue it to the poster board. You should now have a convex mask protruding from the left side of the poster board and a concave mask indented into the right side, as shown in figure 1.

To Do and Notice

Stand the board up on a table by propping it against a box, a stack of books, or anything convenient. Place the light source a little less than 3 feet (1 m) from the concave mask and a little off to the side.

With your head about 3 feet (1 m) away from the masks, close one eye and look at the masks. Move your head to the left and right, or walk to the left and right.

Both masks should look convex—as if they are protruding out from the poster board—even though the one on the right is hollow, or concave.

Additionally, the hollow mask should seem to follow you as you move.

Keeping one eye closed, try moving your head up or down while looking at the hollow mask. It should still follow you.

Try moving the light to illuminate the masks from different angles, and try adjusting the intensity of the light.

What's Going On?

It's hard to judge depth with only one eye. When you see the masks with one eye closed and from some distance away, your brain assumes that both masks protrude outward, as ordinary faces do.

But why does the concave mask seem to follow you when you move? This illusion is your brain's attempt to make sense of two conflicting sources of information. First, your visual system notices that the nose of the concave mask moves less than the rest of the face when you move your head. This information suggests that the nose is the most distant feature on the face (and the face is therefore concave), because, to a moving observer, distant objects appear to move less than nearby objects. (Think of how highway signs whiz by when you are in a car, while the distant scenery

Figure 2 *Does one of these two circles appear to protrude from the page like a bump? Does the other one look more like a crater? Hold the book upside down and the situation will probably change.*

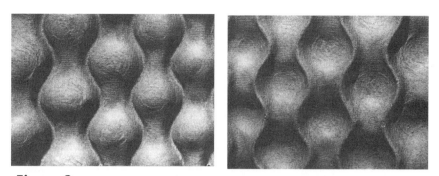

Figure 3 *The left image shows an egg crate, illuminated from above. The right image shows the same egg crate illuminated from below. Most people see protrusions in the left image and indentations in the right image. Hold the book upside down and look again.*

seems to barely move at all.) But based on its experience in the world, your brain can't accept the existence of a concave face. So, it concludes that the hollow mask is instead an ordinary protruding face that turns to watch you walk past, its nose following your gaze and therefore having little apparent motion.

The intensity and angle of illumination can influence the effectiveness of the illusion. The areas of shadow and bright reflection can either enhance or detract from the illusion, depending on their location.

So What?

Disneyland's Haunted Mansion uses hollow masks to create eerie illusions of faces that seem to rotate and follow you as you move. The placement of the masks and the lighting are carefully adjusted to maximize the illusion.

Did You Know?

Researchers in visual perception have traditionally concluded that, in the absence of other cues, we normally assume objects are illuminated from above (based on the fact that the sun is such a basic source of illumination). We therefore perceive objects with shadows at the bottom as convex and those with shadows at the top as concave.

Nonetheless, we tend to perceive all human faces as convex—even hollow masks that have shadows at the top. Our perception of faces seemed to be an exception to our usual interpretation of shape based on light and shadow and was ascribed to the special importance that faces have for us.

However, neuropsychologist V. S. Ramachandran has found that this propensity to view hollow masks as projecting outward actually extends to many complex inward-curving objects. His work generalizes this tendency, rather than attributing it to a unique perception of the human face by the brain.

Going Further

Painted Faces

Try painting the features on the inside of a mask to create your own realistic Haunted Mansion–type mask (see So What?). Or just try looking at the inside of a Halloween mask with a reasonably bright light source behind it, so that the painted features become visible through the mask.

Credits & References

This snack is based on the Exploratorium exhibit of the same name.

Brand, Judith. "How We Learn by Being Fooled: The Lessons in Illusions." *Exploring* (Visual Illusions issue), Vol. 20, No. 2, Summer 1996. *Exploring* is now called *Exploratorium Magazine* and is the museum's quarterly magazine. This article is an outstanding discussion of visual illusions in general, with examples of several types. The work of V. S. Ramachandran in exploring our tendency to perceive complex concave objects as convex is discussed on page 11.

Gregory, Richard. *Eye and Brain: The Psychology of Seeing.* 4th ed. Princeton, N.J.: Princeton University Press, 1990. See pages 190–195 for a discussion of the role of shading and shadow in the perception of concavity and convexity, including the egg-crate illusion shown here in figure 3.

Gregory, Richard. *The Intelligent Eye.* New York: McGraw-Hill, 1970. A discussion of the hollow mask illusion appears on pages 126–131.

Yellott, John, "Binocular Depth Inversion." *Scientific American,* July 1981, pages 147–159. Although this article deals primarily with binocular (viewing with two eyes) inversion, monocular inversion with a hollow mask is discussed, and the hollow mask is part of an interesting experiment described in the article.

Saltwater Pentacell

Current events in electrochemistry.

Make your own battery! Create five simple cells from aluminum foil, copper wire, and salt water, and connect them in a series. Together, they produce enough current to light an LED.

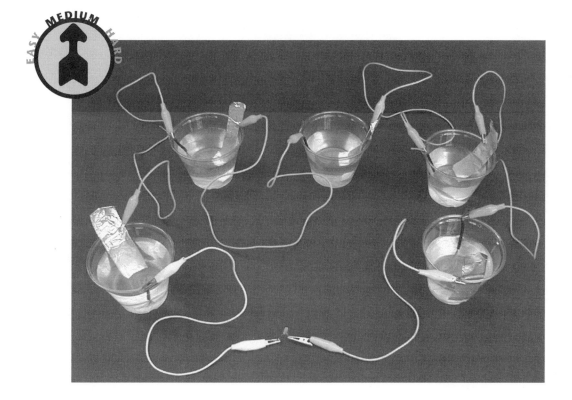

Materials

- scissors
- 20 in (50 cm) of stranded copper wire, 18 or 20 gauge; ordinary lamp cord works well—just split the two strands of a 10-in (25-cm) piece
- wire strippers
- aluminum foil, about 8 in (20 cm) from a normal 12-in-wide (30-cm) kitchen roll

- table salt (sodium chloride), about 2 tablespoons (30 mL)
- water, about 1 quart (1 L)
- pitcher or bowl with a spout
- 5 plastic cups
- 6 alligator-clip leads (e.g., RadioShack #278-1156 or #278-1157; the latter have heavier-gauge wire)

- light-emitting diode (LED); RadioShack #276-330 is an inexpensive LED that works very well for this application (some LEDs glow a little more brightly than others, but most seem to work satisfactorily)
- vinegar (acetic acid), about 1 tablespoon (15 mL); possibly optional—see Helpful Hint

1 Cut the wire into five sections of 4 inches (10 cm) each. Strip 2 inches (5 cm) of insulation off one end of each of the five pieces, and then strip 1 inch (2.5 cm) of insulation off the other end of each piece. This will leave a 1-inch (2.5-cm) piece of insulation on each piece that will act as a sleeve holding the bundle of fine wires together. Twist the strands at the 1-inch (2.5-cm) end of each piece tightly together. Then separate the strands of each 2-inch (5-cm) end so that the loose strands look something like a broom (see figure 1). These are your copper electrodes.

Figure 1

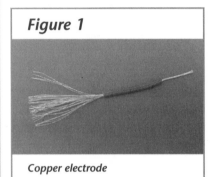

Copper electrode

Figure 2

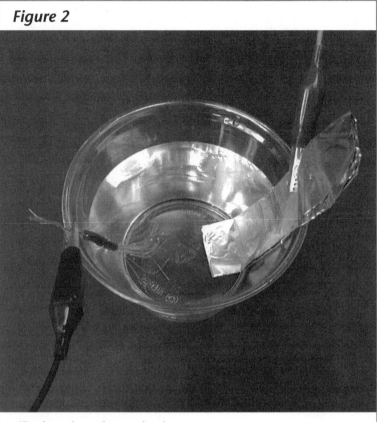

Clip electrodes to the top of each cup.

2 Cut five pieces of aluminum foil, each about 4 inches × 4 inches (10 cm × 10 cm). Fold each piece in half, and then again in half parallel to the first fold, so that it ends up four layers thick, with final dimensions approximately 1 inch × 4 inches (2.5 cm × 10 cm). These are your aluminum electrodes.

3 Add the salt to the water and stir. This is the *electrolyte solution*—a liquid that can conduct electricity.

4 Fill each cup about three-quarters full of the electrolyte so-lution. Then put one aluminum electrode and one copper electrode in each cup. The broomlike end of the copper electrode should be in the solution.

5 Each cup and its electrodes make up one saltwater cell. Connect the cells in series by clipping alligator-clip leads from the copper electrode of one cup to the aluminum electrode of the next cup, and so on, until all five cells are connected (see the opening photo). As you attach each alligator clip to an electrode, you can simultaneously clip the electrode to the top of the cup to hold it in place as shown in figure 2. When you are done, the aluminum electrode in the first cup and the copper electrode in the fifth cup should be left unconnected.

6 Adjust the two electrodes inside each cup as necessary to make sure that they do not touch each other.

7 Use alligator-clip leads to connect the aluminum electrode in the first cup to one leg of the LED and the copper electrode in the fifth cup to the other leg.

To Do and Notice

Did the LED light? Sometimes you have to look directly into the end of an LED to tell if it is lit or not. If you are in doubt, darken the room or cup your hands around the LED to block the room light. If the LED isn't lit, reverse the legs. (A diode—in this case a light-emitting diode, or LED—allows current to flow in only one direction. If it's connected "backwards," it won't light.) If the LED still doesn't light, add vinegar to each cup (see Helpful Hint).

Try using four cells to light the LED. If it lights, then try using three cells. What is the smallest number of cells that will do the job?

Helpful Hint

If the LED won't light after you have followed the normal procedure, try adding a half teaspoon (2.5 mL) of vinegar to each cup and stirring. The acidity of water varies from place to place, and if your water is not acidic enough, the vinegar may make a difference. If your water is already acidic enough, you probably won't need to use the vinegar.

What's Going On?

Each cup, with its electrodes and the electrolyte solution, is a simple electrochemical cell. The two electrodes are made of dissimilar materials (in this case, two different metals) with different chemical activities. A tug-of-war for electrons occurs between the two electrodes, resulting in a potential difference, or voltage. In the cells you have made, aluminum is the more active metal—atoms of aluminum lose their electrons more easily than do atoms of copper. The potential difference causes electrons lost by the atoms in the aluminum electrode to travel through the LED to the copper electrode, and this flow of electrons is the electric current that lights the LED.

DOCTOR FUN
FARLEY

"Galvani! Stop playing with those batteries and eat your frog legs!"

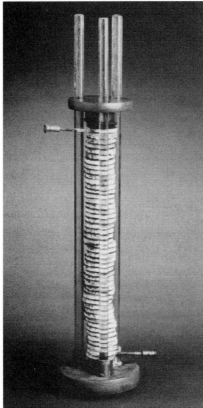

Each layer in this stack is an electrochemical cell. The entire pile is a battery, known as a voltaic pile.

If this flow of electrons continued, and nothing else happened, then fairly quickly there would be a buildup of electrons on the copper electrode and a shortage of electrons on the aluminum electrode.

Because electrons have a negative charge, this would result in the copper electrode becoming negatively charged and the aluminum electrode becoming positively charged. Additional electrons that tried to move from the aluminum to the copper would be repelled by the copper and attracted back to the aluminum, and electron flow would stop.

This is where the saltwater electrolyte solution comes into play. Salt is sodium chloride, and when it's dissolved in water, it forms positive sodium ions and negative chloride ions. The positive sodium ions are attracted to the negative copper electrode, where they participate in neutralizing the extra negative charge through chemical reactions. Likewise, the negative chloride ions are attracted to the positive aluminum electrode, where they participate in neutralizing the extra positive charge. Therefore, there's a constant flow of charge from one electrode through the LED to the other electrode and then through the electrolyte solution, forming a complete circuit.

The five cells make up a *battery* when they are connected in series. (A battery is two or more electric cells that are joined together.) The five-cell battery has five times the voltage of each individual cell.

The chemical reactions that occur at the electrodes of a simple cell can be summarized in chemical equations. The equations for some simple cells can be found in high school and college chemistry texts. The detailed chemistry of some batteries, however, is quite complicated, and in many cases is not completely known.

It takes a minimum voltage to light an LED. If you don't have enough cells, you won't provide the necessary voltage.

So What?

The big idea in this very simple battery is the difference in the abilities of two materials to lose and gain electrons. This same idea is at the heart of the wide variety of batteries used for everything from flashlights to digital watches. The materials, size, and shape of these batteries may differ from those of this saltwater pentacell, but the general principle remains the same.

Did You Know?

Voltage and Current Are Different!

In any electrochemical cell, the greater the difference in the activity of the two materials making up the electrodes, the greater the strength (voltage) of the cell. The larger the size of the electrodes, the greater the number of electrons that can be pulled per second and the larger the current (measured in amperes, or amps).

Going Further

A Solid Choice?

Why does this snack use stranded wire with the wires spread apart? Try substituting a piece of 18- or 20-gauge solid copper wire. Do you get the same results?

Metals Are Not Created Equal

Try using other metals for electrodes. Can you find metals that will allow you to light the LED using fewer cells? Galvanized nails can be used for zinc, regular iron nails for iron, old silverware for silver, and brass hardware for brass. (A commonly available nonmetal that can also act as an electrode is carbon pencil lead.)

Measure It

If you have an electrical meter available, try making quantitative measurements of voltage and current for different combinations of metals.

Make a Buzz

Try substituting a 1- to 3-volt piezoelectric buzzer for the LED (e.g., RadioShack #273-065 or #273-075).

Credits & References

We were introduced to this snack by Art Morrill.

Kluger-Bell, Barry. "Pickle Power." *Exploratorium Quarterly* (Electricity Issue) Vol. 14, No. 3, Fall 1990, pages 25–29. This article describes the author's experiments with homemade batteries that use easily obtainable household materials. Also included are instructions for making a simple ammeter to detect electric current.

Shakashiri, Bassam. *Chemical Demonstrations: A Handbook for Teachers of Chemistry*, Vol. 4. Madison: University of Wisconsin Press, 1992. See pages 91–95 for a general discussion of batteries.

Sensitive Filament

Use your breath to create a visible change in an electric circuit.

The exposed filament from a 100-watt light bulb is wired in series with a flashlight bulb and a 9-volt battery. When you blow on the exposed filament, the flashlight bulb gets brighter.

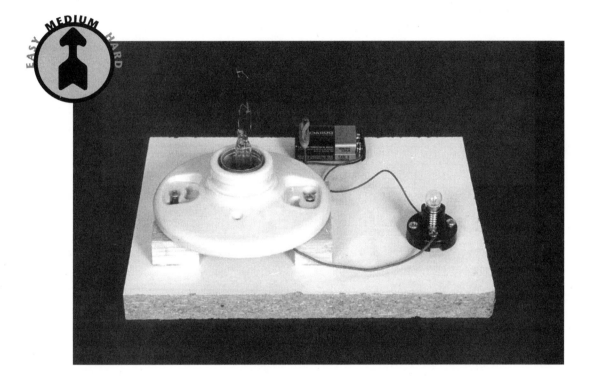

Materials

- snap-type battery clip (e.g., RadioShack #270-324 or #270-325)
- standard light bulb socket
- threaded socket for flashlight bulb (e.g., RadioShack #272-357)
- wire stripper
- insulated wire, approximately 20 or 22 gauge, about 8 in (20 cm) long
- mounting board, approximately 8 in × 12 in (20 cm × 30 cm); the board just has to be big enough to mount the components
- drill
- drill bits of appropriate size for screws
- screwdriver
- 2 medium wood screws to mount large socket
- 2 small wood screws to mount small socket
- 9-volt battery
- 2 nails, $\frac{3}{4}$ in or 1 in
- rubber band
- 100-watt light bulb (filament intact)
- plastic or paper grocery bag
- hammer
- needle-nose pliers
- 3-volt flashlight bulb with threaded base (e.g., Radio Shack #272-1124 or #272-1132)

1 Connect one of the battery clip's wire leads to the socket, and the other lead to the small socket.

2 Strip the insulation off the ends of the piece of wire. Connect one bare end to the remaining contact of the larger socket, and the other end to the remaining contact of the small socket.

3 Place the two sockets on the board and drill pilot holes into the board through the mounting holes of the sockets. Attach the sockets to the board with the wood screws.

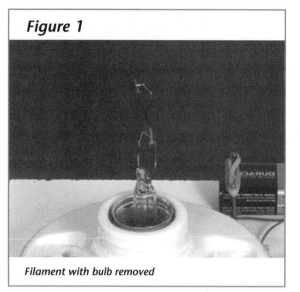

Figure 1

Filament with bulb removed

4 Connect the battery to the battery clip and place the battery on the board. Drive the nails into the board on either side of the battery. Wrap the rubber band over the two nails to hold the battery in place.

5 Now you must remove the glass from the 100-watt light bulb without breaking the filament. Put the bulb into the grocery bag and place the bulb and bag on a hard surface. Tap the bulb with the hammer just hard enough to break the glass. Remove the bulb *carefully* from the bag—the filament of the bulb is very fragile and can break if it touches pieces of shattered glass. Use needle-nose pliers to remove any glass fragments remaining around the base of the bulb.

6 Screw the exposed filament into its socket on the mounting board (see figure 1).

7 Screw the flashlight bulb into its socket. The circuit is now complete, and the flashlight bulb should light up.

➡ Helpful Hints

If you break the filament on a new bulb while breaking and removing the glass, you will have to throw the bulb away and start over. To minimize the chances of this, it's a good idea to practice breaking the glass on a couple of burned-out bulbs first. Any wattage will do. After you get the feel of the process, try a new bulb.

If you try step 5 and don't succeed, here's another method to try. Wrap the glass with packaging tape and use a sharp file to cut the glass bulb at its base. Then carefully lift off the glass bulb.

Perhaps you noticed that the large socket in the introductory photo is raised on two small wooden blocks (the blocks are hot-glued to the base). It was built this way so that the wire connections on the bottom of the socket could be seen. We haven't specified this in our Assembly instructions, and we're drawing it to your attention for two reasons. First, we wanted to avoid confusion in the event that you noticed it and wondered why we didn't mention it in the instructions. Second, we wanted to make the point that sockets will differ (some have connections on top and some underneath), but the exact socket model, and the exact location of things on the base, are not really crucial. It is important, however, that the sockets and the battery are wired in series.

To Do and Notice

Blow gently on the exposed filament, and notice any change in the brightness of the flashlight bulb.

Before you read What's Going On?, try to figure out why blowing on the filament affects the brightness of the flashlight bulb.

Here's one experiment to try: Unscrew the flashlight bulb until it goes off, and let the exposed filament cool off for about 15 seconds. Then screw the flashlight bulb back in and observe its brightness very carefully for the first second or two after it lights.

What's Going On?

When you blow on the exposed filament, you cool it off because the air current carries away a fair amount of heat energy. As the temperature of the filament decreases, its electrical resistance decreases as well. This is because the atoms making up the filament vibrate less at lower temperatures, making collisions between the atoms and the electrons moving through the filament less likely. With fewer collisions, the electrons move more freely through the filament—in other words, they encounter less resistance.

Lowering the resistance of the exposed filament lowers the resistance of the complete circuit, allowing the flow of current in the circuit to increase. Since the flashlight bulb is part of the complete circuit, current through it

also increases, making it glow more brightly.

When you screw in the flashlight bulb, its tiny filament heats and glows almost instantaneously, but it takes the large exposed filament a second or two to reach maximum temperature. For the short amount of time that the large exposed filament is relatively cool and has low resistance, the flashlight bulb glows very brightly; but once the exposed filament heats up and its resistance increases, the current in the complete circuit is reduced and the flashlight bulb dims.

So What?

When you turn on an incandescent lamp, the filament starts out at room temperature. While the filament is relatively cold, it has a low resistance; it draws a large pulse of electric current at first, then settles down to a lower constant current. The initial burst of current is ten times greater than the constant current. That's why light bulbs tend to burn out when they're first turned on: The initial large rush of current causes stress in the filament.

Blow on the exposed filament and watch the companion light bulb brighten.

Going Further

Heating and Cooling

Use other means to change the temperature of the exposed filament (a match flame, ice cubes, a hair dryer, etc.), and observe any change in the brightness of the flashlight bulb. Can you explain the changes that occurred? **CAUTION:** The exposed filament can get uncomfortably hot and is extremely fragile.

Credits

We first saw this activity done by the Galileo Circle, a group of Japanese science teachers, and we would like to acknowledge their contribution.

Shadow Panel

In a flash, make a shadow with a life of its own.

Normally, the life of your shadow is at your mercy. If you move, your shadow moves. In this snack, you use a luminous screen and a disposable flash camera to create a shadow that hangs around for awhile.

Materials

- 1 sheet of Glow Max phosphor-coated glow-in-the-dark paper (available in some office supply, stationery, and art supply stores; contact Riverside Paper Co., 920-991-2210, www.riversidepaper.com, for the

nearest retail outlet). **NOTE:** If you have trouble finding this paper, you can make good shadow panels using poster board or foam core and luminous paint (see Alternative Construction).

- used disposable flash camera without film (Ask a one-hour photo facility to save some used disposable cameras for you; many places will do this for teachers and students at no cost.)

1 Tape a sheet of Glow Max paper to a wall, or lay it on a table. This is your luminescent shadow panel. Make sure the room can be fully darkened.

2 Figure out how to set off the flash on the camera. Different brands work differently. If you have any trouble, follow these steps:

a. Remove the camera back if it's not already off (a little prying will be necessary).

CAUTION: If removing the back exposes any electronic parts (other than the battery), be careful not to touch them. They are capable of giving a nasty shock. (If the front of the camera comes off also, exposing electronic components, tape it back on with masking tape, so that the electronic parts are covered.)

b. With the back of the camera removed, you should be able to see a portion of the sprocket wheel, which has teeth that fit in the holes along the edge of the film when the camera has film in it. Turn the sprocket wheel slowly until you feel it click into a cocked position. (The sprocket wheel should not be confused with the serrated wheel used to advance the film. This latter wheel is visible even when the camera back is still on and is now useless with no film in the camera.)

c. Be sure there is a working AA battery in the camera and that it is oriented properly. Activate the flash-charging button until you see the flash indicator light glow. (The flash-charging button is normally located on the front face of the camera, and the flash indicator light is usually located on the back of the camera near the viewfinder window.) Press the shutter button to set off the flash. (If you can't get the flash to work, try advancing the sprocket wheel a little past where you first think it is cocked.)

Alternative Construction

Luminous paint brushed onto white poster board or foam core works just as well as glow-in-the-dark paper. Luminous paint, also known as "glow-in-the-dark" paint, may be found at art supply stores, craft stores, hobby shops, stationery stores, novelty stores, and at art supply stores on the Internet. Some specific brands are listed below:

- Edmund Scientific Luminous Acrylic Paint (#V31-806, 2-ounce bottle), 800-728-6999, www.edsci.com

- Golden brand Acrylic Phosphorescent Green Glow in the Dark Acrylic Paint (Phosphorescent Medium, #4900, various sizes), 510-649-4800. For locations of stores that carry this product, see www.goldenpaints.com.

- Palmer Luminous Paint (#628612). This brand is available in 2-ounce quantities.

Depending on the consistency of the individual paint, an ounce of luminous paint will cover from two to five $8\frac{1}{2}$-×-11-in pieces of foam core or poster board.

To Do and Notice

Cock the camera so that it is ready to take a flash picture. Darken the room and wait about ten seconds so that the paper's glow dims. Place your open hand (or an object of your choosing) on the luminescent panel or very close to it. Hold the camera about 12 inches (30 cm) away from the panel, point the camera at the panel, and press the shutter button to set off the flash.

Take your hand away or remove the object.

You should see a black shadow image on the luminescent green background of the panel. The shadow should persist for several seconds before fading.

What's Going On?

The glow-in-the-dark paper contains a phosphorescent material. When the electrons in a phosphorescent material absorb light, they jump up to higher energy levels. When they fall back to their original energy levels, they release the energy they absorbed in the form of light (see figure 1).

The flash is a burst of light that reaches the glow-in-the-dark paper only where it is not obscured by the hand (or other object) in front of it. The light is absorbed by the phosphorescent material in the paper and then released over a period of time as a greenish glow. The glow lasts for a while because not all the electrons make the transition back to lower energy levels at the same time.

Darkening the room for a brief time before setting off the flash allows most of the electrons that absorbed energy from the light in the room to fall back to their ground energy states. If you don't let the luminescent panel sit in the dark, even the area where your hand is placed will still be glowing from the room light, and the contrast between the shadow area and the background will not be so great.

So What?

Phosphorescent pigment is used in everyday life, sometimes for fun and sometimes to serve a purpose. There are glow-in-the-dark stars that can be stuck to the walls or the ceiling to create a nighttime sky in your bedroom. Inflatable "star balls" absorb the light of the sun when you play with them during the day, and at night, you can enjoy their luminescent green glow. Some clocks and some watches have phosphorescent pigment-coated hands that glow in the dark.

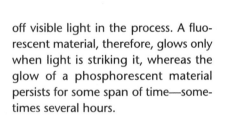

Figure 1 *This energy-level diagram shows electrons excited to higher energy levels, then falling back down to release light.*

Did You Know?

Another Kind of Glow

The phenomenon called *fluorescence* is similar to phosphorescence. Fluorescence occurs when electrons are energized, usually by ultraviolet light, but return to their original energy levels immediately, giving off visible light in the process. A fluorescent material, therefore, glows only when light is striking it, whereas the glow of a phosphorescent material persists for some span of time—sometimes several hours.

Going Further

In a Box

Try making a shadow box that works in a lighted room. Tape a luminous panel to an inside wall of a cardboard box, place an object in the box up against the screen, put the lid on the box, cut a small hole in the wall of the box opposite the screen, snap the flash through the hole, and then open the lid a bit to view the shadow.

Credits

This snack is based on the Exploratorium exhibit Shadow Box.

The Shadow Box exhibit at the Exploratorium is one of the museum's most popular exhibits.

Snip and Snap

Get more bounce per ounce.

If you dropped a ball and it bounced higher than its original height, would you be surprised? In this snack, you will do exactly that, using a cut-up racquet ball. Does this disprove the principle of conservation of energy?

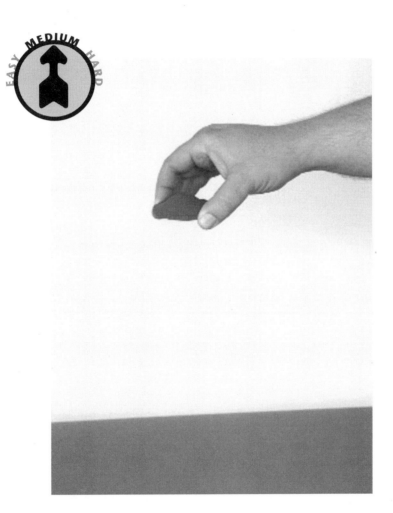

Materials

- racquet ball (used racquet balls can often be obtained free from racquet ball clubs after they have tournaments; unfortunately, a tennis ball won't work in place of a racquet ball)
- hacksaw blade or serrated kitchen knife

- sharp scissors
- Super Ball or other similar high-bouncing ball
- tennis ball (or a second racquet ball, if that's easier to obtain)
- Optional: set of two Rebound balls (one ball in the set is "live" and behaves much like a Super

Ball; the other is totally "dead" and doesn't bounce at all. You can get Rebound balls from scientific supply companies, museum stores, nature stores, etc. If you can't get a set, just ignore the parts of the activity that refer to them.)

1 Find the seam on the racquet ball. (You may have to look closely to find it; sometimes it's hard to see.)

2 Use the hacksaw blade or serrated kitchen knife to start a cut along the seam. Cut in one place until you have cut a small slit completely through the rubber shell of the ball.

3 Insert one of the points of the scissors through the slit, and finish cutting the ball in half along the seam.

4 Place your fingers along the outside edge of the one of the ball halves, and use your thumbs to push the half inside out as shown in figure 1. The ball-half will remain inside out when you remove your thumbs.

5 Trim approximately $\frac{1}{4}$ inch (0.6 cm) off the rim of the ball-half (see figure 2). Your goal is to trim enough off so that the ball-half remains inside out—but just barely. If the ball-half won't stay inside out, then you've cut too much off. Throw that ball-half away and start on the other half, but trim less off initially.

6 Hold the trimmed, inside-out ball-half about $2\frac{1}{2}$ feet (0.75 m) above a flat table, with the round side up as shown in figure 3. Now drop it so that it lands on the table in this same position. It should pop up from the table top and bounce much higher than the height from which you dropped it. If it doesn't, carefully trim a little more off the edge and try again. Keep on trimming and testing, as needed, to get the ball-half to bounce up from the table top. (Remember, if you trim so much that you can't get the ball-half to remain inside-out, you'll need to start over with the other half of the ball.)

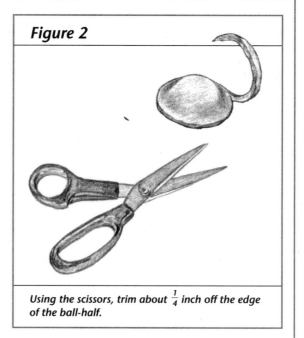

Figure 2

Using the scissors, trim about $\frac{1}{4}$ inch off the edge of the ball-half.

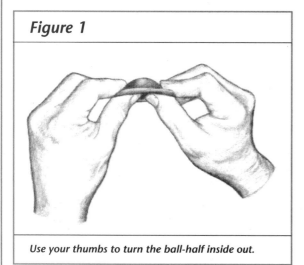

Figure 1

Use your thumbs to turn the ball-half inside out.

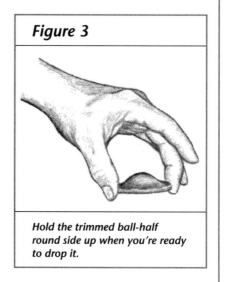

Figure 3

Hold the trimmed ball-half round side up when you're ready to drop it.

To Do and Notice

Drop the Super Ball onto the tabletop and notice how far the ball bounces back up.

Drop the tennis ball (or racquet ball) from the same height, and compare its bounce with that of the Super Ball.

Drop the "dead" Rebound ball from the same height and notice that it doesn't bounce at all.

Finally, drop the inside-out ball-half from the same height. Notice that it bounces far above its starting point.

What's Going On?

When you hold a ball above a tabletop, it has *gravitational potential energy.* Let the ball drop, and that gravitational potential energy becomes kinetic energy as the ball falls. As the ball smashes into the tabletop, the kinetic energy is transformed into the *energy of deformation*—the ball flattens a little when it hits the table. (This happens too fast for you to see it, but high-speed photos show that a ball distorts at the point of impact.)

The flattened ball stores potential energy in the same way that a compressed spring does. When the squashed ball returns to its original shape, the potential energy that it contained becomes kinetic energy again as the ball bounces upwards.

But not all the energy returns to kinetic energy. In the process of deformation, some energy becomes thermal energy, warming the ball. How much of the energy becomes thermal energy depends on what the ball is made of. The Super Ball transforms the least energy to heat; the tennis ball transforms more energy to heat, and the dead Rebound ball transforms nearly all its energy to heat.

Because some energy is lost to heat, none of these balls bounce to the height from which they were dropped. This behavior is what you would expect; it's consistent with the law of conservation of energy.

Things work differently for the inside-out ball-half. It transforms much of its kinetic energy to heat after impact, just like the other balls. Yet, the ball-half bounces higher than the point from which it was dropped. You might think it was defying the law of conservation of energy because it has more gravitational potential energy at the height of its bounce than it did before it was dropped. Where does this energy come from?

That energy comes from the work you did when you turned the ball inside-out. The energy is stored as potential energy, just like the energy in a compressed spring. Hitting the tabletop triggers the release of this stored energy, which sends the ball flying!

So What?

For every sport, the ball is designed to provide just enough bounce to make the game interesting. When dropped, baseballs do not bounce as high as tennis balls. Play a game of baseball using tennis balls, and the game changes completely.

Going Further

Read the Label

If you are able to obtain the Rebound balls, read the description of their molecular structure, which determines their behavior.

Another Experiment

You can do another dramatic ball-bouncing experiment with a basketball and a tennis ball. Hold the tennis ball on top of the basketball and drop them together. The tennis ball will bounce off the basketball and fly off over your head.

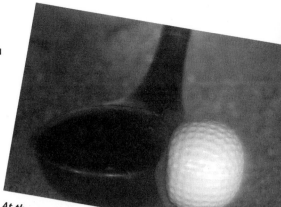

At the moment of impact, the club deforms the golf ball, squashing it flat. When the ball springs back to its original shape, it pushes off the club and the potential energy stored in the ball's deformation becomes kinetic energy as the ball flies through the air.

To get a feel for what's going on here, drop the basketball on its own and notice how high it bounces. Then drop the basketball and the tennis ball together and notice how high the basketball bounces. When you send the tennis ball flying, the basketball doesn't bounce as high.

How high the basketball bounces is an indication of how much kinetic energy the basketball has as it leaves the ground. When the basketball sends the tennis ball flying, the basketball doesn't bounce as high because it has transferred some of its energy to the tennis ball.

Credits & References

This snack is a low-cost, home-made version of the commercially available Hopper Popper, made by Dynatoy International Inc., Salt Lake City, Utah.

Davis, Susan, Sally Stephens, and the Exploratorium. *The Sporting Life.* New York: Henry Holt and Company, 1997. You'll find more about bouncing balls beginning on page 89. (Some activities from *The Sporting Life* are also available on the Exploratorium Web site, www.exploratorium.edu.)

Soap Film in a Can

What light from yonder soap film reflects?

You've probably seen the iridescent colors that appear in a soap bubble when the light strikes it just right. This effect—the result of interactions between light waves and the soap film—can be seen very clearly when you put the soap film in the opening of a black film canister. The colors of the soap film are bright against the black background of the can.

Materials

- dishwashing liquid
- water
- pitcher or bowl with a spout
- shallow dish
- sheet of white paper
- black film canister without lid
- pencil

1 Make a bubble solution by mixing 1 part dishwashing liquid in 16 parts water. This is equivalent to $\frac{1}{4}$ cup detergent in 1 quart of water (about 60 mL of detergent in 1 L of water). If you want to make a larger quantity, use 1 cup (240 mL) detergent in 1 (4 L) gallon of water.

2 Fill a shallow dish approximately a half inch (1 cm) deep with the bubble solution.

3 Place the white paper on a tabletop.

To Do and Notice

Dip the open mouth of the film canister into the soap solution, and then pull it out.

In a brightly lit place, hold the canister horizontally about an inch (a few centimeters) over the white paper, so that the soap film is in a vertical plane.

Watch the colors form and move in the film. Notice the horizontal bands of color.

Notice that after awhile the top of the film becomes "invisible." Look carefully through this part of the film into the canister. What do you see?

Poke a pencil point into this invisible region of the film. What happens?

What's Going On?

The soap film is a water sandwich. A layer of water is held between two layers of soap molecules. When the soap film is vertical, the water slowly drains down under the pull of gravity, thinning the top part of the film and thickening the bottom part.

Most of the light striking the soap film passes through the film, but about 4 percent reflects from the front surface of the film back into the air, and another 4 percent or so reflects from the back surface of the film. The light waves reflecting from the two surfaces interfere with each other in different ways, depending on the wavelength of the light and the distance between the front and back surfaces of the film.

For each wavelength of light, there are soap-film thicknesses that will put the two light waves *in phase*, with their crests and troughs lined up as shown in figure 1. These waves

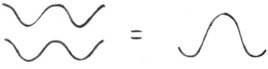

Figure 1 *Constructive interference*
Combined, these waves reinforce each other.

interfere constructively—they reinforce each other, and the color that corresponds to their wavelength is made more strongly visible. For each wavelength of light there are also soap-film thicknesses that will put the two light waves *out of phase*, with the crests of one lined up with the troughs of the other (see figure 2). These waves

Figure 2 *Destructive interference*
Combined, these waves cancel each other.

interfere destructively—they cancel each other out, and the color that corresponds to their wavelength is removed from the light entering your eyes.

You perceive bands of color on the soap film because at each location on the film a different color of light is being destructively removed. Where the thickness of the soap film causes the destructive removal of one of the primary colors that makes up white light, you see a mixture of the two remaining colors:

white – red = blue + green = cyan (bluish green)

white – green = red + blue = magenta (reddish blue)

white – blue = red + green = yellow

What's causing each specific color you perceive in different parts of the soap film? Let's start with the top part of the film, which should have become invisible. The invisibility of this part of the film is the result of all the light reflecting from it getting canceled out by destructive interference. This occurs because the soap film is thinner than any of the wavelengths of visible light. The back part of the film is so close to the front part that the waves reflecting off the two are barely shifted in relation to each other.

At the same time, it happens that the light waves reflecting from the front of the soap film are inverted— dumped upside down—while those from the back are not. As a result, all the wavelengths of light reflecting from the front of the film are exactly out of phase with the wavelengths of light reflecting from the back of the film, and they all cancel out, as shown in figure 3 on the next page.

Below the invisible portion of the film, you should have noticed a silver-colored region. In this region, the film is about one-quarter of a wavelength of blue light in thickness. For this

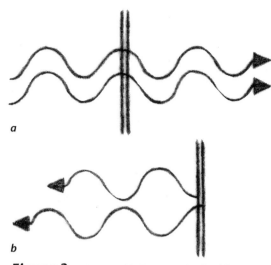

Figure 3 *Behavior of light waves in the thinnest part of a soap film. Most of the light passes through the film (a), but some reflects from the front surface and some from the back surface (b). The light waves that reflect from the front surface are inverted (turned upside down) and become out of phase with the waves reflecting from the back surface; as a result, they cancel each other when they combine.*

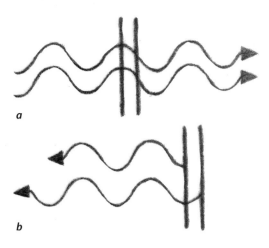

Figure 4 *Behavior of light waves in a soap film that is one-quarter of a wavelength thick. Most of the light passes through the film (a), but some reflects from both the front and back surfaces (b). The wave reflecting from the back surface must travel an extra one-half wavelength—one-quarter of a wavelength in each direction—before it reenters the air in front of the film. The wave reflecting from the front surface is inverted. These two effects put the waves in phase, and they reinforce each other when combined.*

thickness of soap film, the blue light waves reflecting from both the front and back surfaces of the soap film are able to add together to produce a strong blue color (see figure 4). All other colors are also reflected, although somewhat more weakly. The sum of all these reflected colors is bluish-silver.

Where the film is one-half of a wavelength of blue light thick, the blue waves cancel. (**Note:** The wavelength of light in the soap film is shorter than the wavelength in air by about 30 percent.) But at this location the film is also one-quarter of a wavelength of red light thick (red light has a wavelength about twice that of blue light). Here, the red waves are reinforced by constructive interference for the same reason that blue waves are reinforced where the film is one-quarter of a wavelength of blue light thick. The result of canceled blue light and reinforced red light is a reddish band.

In general, where blue light is strengthened, red light is weakened, and vice versa. The result is alternating bands of color. Red and blue are not the only colors involved; the same thing is happening to other colors (green, yellow, etc.) at each film thickness. The color bands are not pure colors; rather, they are the overall result of the combining of the varying intensities of many colors at each particular thickness of film.

So What?

The coatings applied to camera lenses affect light in much the same way as soap films do. The thin-film optical coating deposited on many camera lenses is chosen to minimize the reflection of light in the middle of the visible spectrum (orange, yellow, and green). The thickness of the coating causes the reflecting waves in these wavelengths to cancel each other. By minimizing reflection from the lens, transmission of light through the lens is maximized, and more light reaches the film in the camera. When reflection is minimized in the middle of the spectrum, however, it is not minimized at the ends of the spectrum (red and blue). Thus the reflected light is relatively richer in red and blue, and the lens looks purple or magenta.

Sometimes you see colors of an oil slick on a puddle in the street. These colors look a lot like the soap film colors you see in this snack. The layer of oil that makes these colors is about the same thickness as a wavelength of light. The light reflecting from the top and bottom of the oil layer combines to create colors in the same way that light reflecting from the soap film does.

Going Further

Soap Film Dome

Drill a hole in the bottom of the film canister with a $\frac{3}{16}$-inch drill bit. Create a soap film over the mouth of the film can, then blow through the hole. The soap film will bulge out into a dome. Then allow the air to rush out of the hole. You can feel the breeze coming from the hole with your hand or see it by holding the hole near a candle flame. Notice the dome flatten out. The pressure in the film can is greatest when the radius of curvature of the dome of soap film is least.

The interesting thing is that the radius of curvature is least when the dome is largest (see figure 5). So the larger the dome, the higher the pressure and the faster the air rushes out of the can. Therefore, the soap film dome deflates rapidly at first and slows down as it flattens. If you plug the hole and place the film can with the opening facing upward surmounted by a soap film dome, colored rings will appear in the soap film.

Did You Know?

Hooke's Mistake

Robert Hooke, a seventeenth-century English scientist, may have been the first to observe a soap film that was thin enough not to reflect light and thus appear invisible. But instead of describing it as invisible in his letter to the Royal Society, he wrote that it appeared not to exist and that some force held the rest of the bubble film in place. You prove Hooke wrong when you poke the invisible part of the film and cause the whole film to break.

The Thickness of Black

Soap bubbles have two different stable thicknesses that look black. The thicker of these is called the common black film. It is 30 nanometers thick—the thickness of about 300 atoms, or 10 soap molecules, or $\frac{1}{20}$ the wavelength of red light. The thinner film is called the Newton black film. It is about 6 nanometers thick—the thickness of two soap molecules, or $\frac{1}{100}$ the wavelength of red light. The Newton black film is much more transparent; that is, much "blacker."

Credits

Linda Hjelle contributed to the development of this snack.

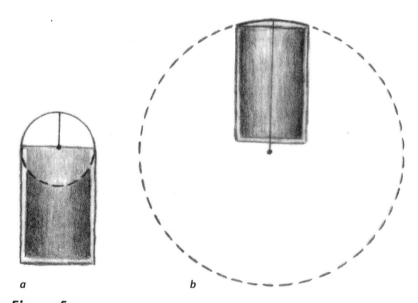

a   b

Figure 5 *The soap film makes a part of the sphere over the mouth of the film canister. The dashed lines show the entire sphere. Note that the soap film portrayed in (b) is part of a larger radius sphere.*

Sound Bite

A new way to listen to music.

When you listen to a radio, you expect to hear the sound coming from its speaker or headphones. But sound doesn't have to be transmitted through the air to be heard. In this snack, you pick up sound vibrations through your teeth!

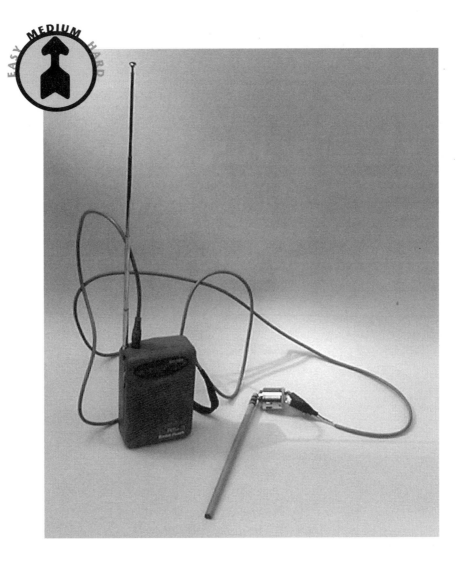

Materials

- pushpin
- new wooden pencil with eraser
- small DC motor (e.g., RadioShack #273-223, 1.5–3 volt)
- small radio with headphone jack (e.g., RadioShack #12-799)

- audio cable, 6 ft (2 m), with a $\frac{1}{8}$-in phone plug (sometimes called a mini plug) on one end and two alligator clips on the other (e.g., RadioShack #42-2421)

- plastic wrap, enough to wrap around the bottom half of a pencil to provide a sanitary covering that you can bite on

1 Use a pushpin to make a hole in the metal jacket of the pencil where it holds the eraser, and then extend it on into the side of the eraser. Wiggle the pushpin around to enlarge the hole in the metal jacket until it is just big enough for the motor shaft to fit in. You don't need to remove any of the eraser material; you are really just creating a channel for the motor shaft to squeeze into.

2 Insert the motor shaft carefully into the hole you have made.

3 Locate a radio station with a clear, strong signal. Leave the radio on.

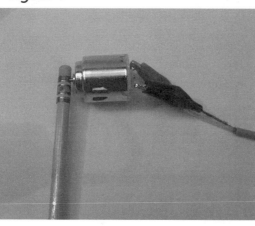

Figure 1

Pencil, motor, and alligator clips

4 Insert the phone plug on the audio cable into the headphone jack on the radio. Connect the alliga-

tor clips at the other end of the cable to the two terminals of the motor. (When you insert the phone plug into the headphone jack, the external speaker is disconnected; the same thing happens when you use headphones.) Figure 1 shows the pencil, motor, and alligator clip assembly.

5 If more than one person will be experimenting with this snack, each person should have his or her own piece of plastic wrap for covering the bottom half of the pencil. If you can find straws big enough to fit over the pencil, you can use them instead of plastic wrap, or you can improvise your own covering.

To Do and Notice

With the radio at medium to high volume, bite down on the pencil. You can either place the bottom end in your mouth as if it were a straw, or put the pencil sideways in your mouth like a dog bone as shown in figure 2. Try both, and use whichever gives you the best results. You should hear the radio playing in your ears! If you have trouble hearing the radio, try plugging your ears with your fingers to drown out any competing external noise. (If you continue to have trouble, see the Helpful Hint.)

Experiment with the following: Try putting the pencil behind your ear against your skull, then hold the pencil against your forehead; notice which position produces the clearest sound. You might also put the motor itself directly against the bones in your head.

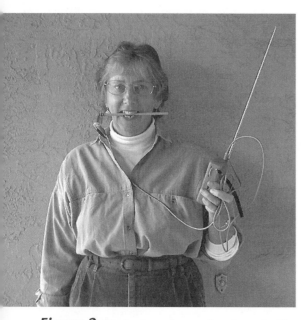

Figure 2 *Bite firmly on the pencil and sound will fill your head.*

What's Going On?

The output from the radio, in the form of a changing electric current, is sent through the audio cable to the motor. When the electric current passes through the coils of the motor, these coils act as electromagnets. Since the

> ➡ **Helpful Hint**
>
> If you can't hear the radio, try drilling a hole through the pencil with a $\frac{1}{16}$-inch drill bit about an inch (2.5 cm) from the eraser and insert the shaft of the motor at that point.
>
> Newer radios that have weaker outputs from the headphone jacks may not work with this snack. If necessary, use older radios or try different radios until you find one that works.

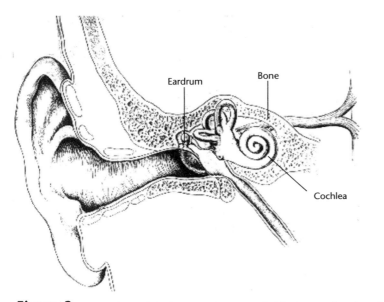

Figure 3 *The cochlea of the inner ear is surrounded by bone. Vibration of the bone vibrates the cochlea, which senses the vibration as sound. Normally, sounds are transmitted from the eardrum to the cochlea by three small bones in the inner ear.*

Did You Know?

Wood, Tooth, and Bone

During the normal hearing process, sound waves spiral through the ridges of your outer ear, bounce around in your ear canal, and vibrate your eardrum as they make their way to the cochlea—and ultimately to your brain. During their journey from the external ear to the cochlea, sounds are amplified more than one hundred times. The sound waves you hear when you bite the pencil, in contrast, are not amplified nearly as much because they are conducted to the cochlea directly through wood, teeth, and bone.

electric current is constantly changing, the strength of the electromagnetic fields is constantly changing also, in synchronization with the radio output. The interaction of these constantly changing electromagnetic fields with the permanent magnets in the motor causes the motor to physically vibrate. The motor vibrations are transmitted to the pencil and then to your teeth and jawbone. Eventually the vibrations stimulate the nerve endings in your cochlea, which is part of your inner ear (see figure 3). The nerve impulses sent by the cochlea along the auditory nerve to your brain are then interpreted as sound, just as if they had been caused by sound waves entering your ear.

So What?

There are two types of hearing loss, conductive and sensorineural. In *conductive hearing loss,* sound vibrations are not being transmitted from the outer ear to the cochlea. In *sensorineural hearing loss,* the brain is not receiving nerve signals from the inner ear, even though sound vibrations may be reaching the cochlea. For people with permanent conductive hearing loss, there are hearing aids that vibrate the bones of the skull. The vibrations travel through other bones until they reach the inner ear.

Hearing aids that rely on bone conduction of sound have been available for awhile. The first one—called an "acoustic fan"—appeared around 1900 (see figure 4). The hard-of-hearing person held the base of the device between his or her teeth and inclined the fan-shaped part toward the source of a sound. As in this snack, the sound vibrations traveled from the teeth to the jawbone and finally to the inner ear.

Figure 4 *The acoustic fan, a hearing aid used in the early 1900s, relied on bone conduction.*

Going Further

Motor Autopsy

If you have never seen the inside of a small electric motor such as the one used in this snack, try taking one apart. Use a broken one if possible, but a new one is not very expensive; its loss is well worth the experience.

Check Your Hearing

If you have access to a tuning fork, you can perform two simple hearing tests, the Rinne Test (pronounced reh-NAY), and the Weber Test, both of which distinguish between conductive and sensorineural hearing loss. Look them up on the Web to learn how to perform them.

Credits & References

The original inspiration for Sound Bite was the Bite-a-Phone in *The Dick and Rae Physics Demo Notebook* (see below). Gabe Espinda, Tien Huynh-Dinh, Eric Kielich, and James Kliewer contributed to the evolution of the present version.

Carpenter, D. Rae Jr., and Richard B. Minnix. *The Dick and Rae Physics Demo Notebook.* Lexington, Va.: Dick and Rae, 1993. A rich resource for teachers, this book contains more than 600 demonstrations from two decades of workshops given by the authors at Virginia Military Institute. Send inquiries to Dick and Rae, Inc., VMI Mallory Hall, Lexington, VA 24450.

Sprotating Cylinder

Create a triangular illusion with a cylinder that's both spinning and rotating.

A piece of pipe with a mark at each end is set spinning and rotating at the same time. We coined the word *sprotating* to describe these two simultaneous motions. Amidst the blur of the moving cylinder, one of the marks appears in three places, forming an apparently stationary triangle.

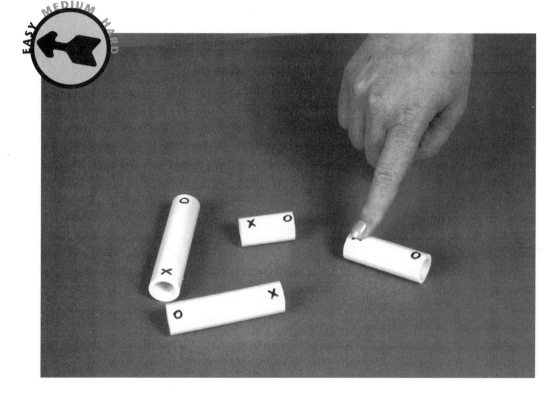

Materials

- PVC shears or hacksaw; inexpensive PVC shears are amazingly handy for cutting PVC pipe and can be bought at hardware or home improvement stores

- $\frac{1}{2}$-in Schedule 40 PVC pipe, about 2 ft (60 cm) long
- colored marker pens (permanent)

- a smooth surface, such as a tabletop
- a transparent tabletop (optional)

1 Cut a piece of pipe that is three times as long as its outside diameter. (Err on the long side; this snack will still work with tubes up to 3.15 diameters long.)

2 Mark an X near one end of the pipe, and put an O at the other end (see figure 1).

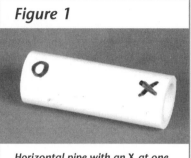

Figure 1

Horizontal pipe with an X at one end and an O at the other

3 After you experiment with the first piece of pipe, you will need a few more pieces. Cut them so that you have a set of pipes with lengths that are two, three, four, and five times their diameters. Put the X and O marks on each pipe.

To Do and Notice

Place the piece of pipe that is three times as long as its diameter on a flat, smooth surface with the marks facing upward. Put your finger on the X, and then rapidly push your finger down while at the same time pulling it toward you (see figure 2).

The cylinder will spin and rotate, making a blurred circle in which Xs can be seen. Notice that as the sprotating cylinder stabilizes, you can see three Xs that mark the vertices of a triangle. Notice that the O does not appear.

Next place your finger on the O and launch the cylinder. Notice that as the motion stabilizes, three Os appear, each at the vertex of a triangle, and the X does not appear.

Try some experiments to figure out what's going on. Here are some suggestions:

- Try the cylinders of different lengths that you've made. Notice the different stable patterns. In particular, try cylinders that are two, three, four, and five times their diameters.

- Make several markings on one end of one of the cylinders.

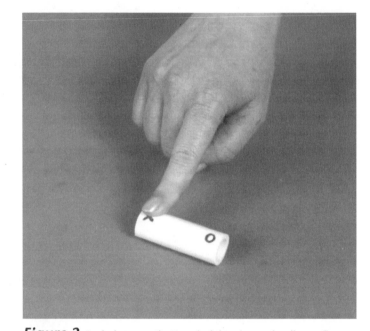

Figure 2 *Push down on the X end of the pipe and pull your finger back toward you at the same time.*

- Look at a sprotating cylinder from underneath, through a transparent table.

- If you've been looking at the cylinder under fluorescent lights, try looking at it in sunlight or under incandescent lights (which do not strobe on and off like fluorescent lights).

- Look at a sprotating cylinder with a stroboscope. (A handmade stroboscope in which slits are cut into the edges of a spinning disk will work just fine.)

- Look at a sprotating cylinder from the side, with your eyes just above the tabletop.

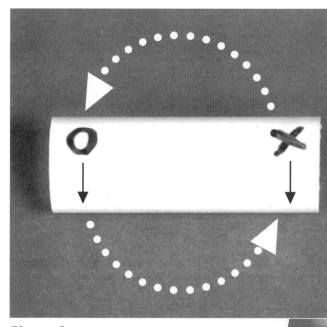

Figure 3 *The dotted arrows show how the pipe is rotating. The solid arrows show how it is spinning.*

What's Going On?

Don't read this until you've done some experimenting!

When you launch the cylinder, it *spins* and it *rotates*. The spinning motion occurs around the long axis of the cylinder, and is shown by the solid arrows on the cylinder in figure 3. The rotation occurs around an axis perpendicular to the table surface, and is shown by the dotted lines and arrows outside the cylinder in figure 3.

As the cylinder sprotates, the top surface of one end (the left end in figure 3) spins in the same direction as that end is rotating, while the top surface of the other end (the right end in figure 3) spins opposite that end's rotation.

On the right end, the two motions cancel each other, so when the mark on the spinning cylinder is facing upward, it actually stops momentarily (see Box o' Math for the explanation). On the left end, the two motions add to each other, so when the mark on

The "Flying Bernoulli Brothers," a pair of wacky physics teachers from Ohio, developed this activity.

the sprotating cylinder comes to the top, it moves twice as fast as it would with either motion alone.

Human eyes can see the stopped mark easily, while the mark moving extra fast is a blur. Thus only the mark on one end is visible.

When you sprotate the cylinder that has a length three times greater than its diameter, you see three stopped marks, and these form a triangle. This behavior means that the cylinder is making three spins for every rotation (see Box o' Math for the explanation).

Why does the sprotating cylinder take a few seconds to settle into a stable triangular pattern? You may have noticed that the cylinder sprotates with one end touching the table and

one end just above the table. Usually the cylinder is launched so that it is spinning faster than it is rotating, which means that the end touching the table slips when it rolls. While this is happening, the pattern isn't stable because the spinning and the rotation aren't constant in relation to each other. As the cylinder rubs against the table, however, energy is quickly dissipated, slowing down the spinning speed until the end that is touching the table can roll without slipping. When this occurs, the pattern formed by the mark on the end of the cylinder becomes stable.

Box o' Math
Three Spins for Every Rotation

Here's a mathematical explanation of why you see three Xs (or three Os) in the sprotating cylinder.

In the discussion below, "right" and "left" refer to the ends of the cylinder in figure 3. Assume that the cylinder has a length (L) which is three times its diameter (d). The right end of the cylinder rotates in a circle with a circumference (C) that is *pi* times the length of the cylinder (L). That circumference (C) could also be described as *pi* times three cylinder diameters (d):

$$C = \pi L$$

Since L is equal to $3d$, then:

$$C = 3\pi d$$

The circumference of the cylinder itself (c) is *pi* times its diameter (d):

$$c = \pi d$$

Divide the circumference of the cylinder's rotation by the circumference of the cylinder, and you get:

$$\frac{C}{c} = \frac{3\pi d}{\pi d} = 3$$

The circumference (C) of the rotating cylinder is three times the circumference of the cylinder (c), so it spins three times for each rotation. If the cylinder begins with the X up, the cylinder will have rolled one third of the way around the circle by the next time the X returns to the top. To complete the rotation, the X must return to the top three times. That's why you see three Xs!

The speed due to rotation (V_r) at either end of the cylinder is equal to the distance the end travels (C) divided by the time (T) it takes for one whole rotation:

$$V_r = \frac{C}{T} = \frac{3\pi d}{T}$$

The speed due to spin (V_s) at either end of the cylinder is equal to the circumference of the cylinder (c) divided by the time it takes for one whole spin (t):

$$V_s = \frac{c}{t} = \frac{\pi d}{t}$$

But there are three spins for every rotation, so $T = 3t$, and therefore:

$$V_r = \frac{3\pi d}{T} = \frac{3\pi d}{3t}$$

$$= \frac{\pi d}{t} = V_s$$

This equation shows that V_r and V_s are equal. This means that for the instant that the marks are facing upward, the mark at the right end actually comes to a stop, because V_r and V_s are in opposite directions. At this point, the mark at the left end moves doubly fast, because the speeds are in the same direction.

A cylinder that has a length four times its diameter will make a square pattern with four marks. A cylinder that is two diameters long will make two marks in a line. A cylinder with a length that is not near an integer multiple of its diameter will make a pattern that is unstable. Such a pattern might, for example, slowly rotate.

So What?

The purpose of this snack is to have fun. You can do simple experiments to figure out how it works, exploring the process of scientific inquiry. Decorate the cylinder in other ways to create works of art and amaze your friends with it at your next party!

Going Further

Try drawing a line down the side of a cylinder. Make one half of the line red and the other half blue. When sprotated, this design produces an artistic three-pointed star. Try spinning a cylinder with a length equal to its diameter. Try different diameter pipes.

Credits

This snack was brought to our attention by Gene Easter and Bill Reitz, also known as the "Flying Bernoulli Brothers."

Square Wheels

You may not be able to put a square peg in a round hole, but you can make a square wheel roll on a round road.

A square wheel will roll smoothly, with its axle at a constant height, on a surface with properly spaced bumps of the right size and shape.

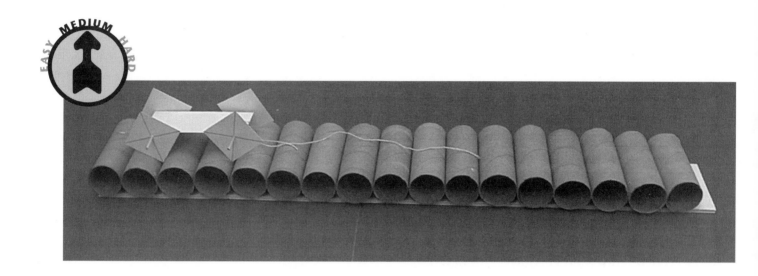

Materials

- hot glue gun and glue sticks
- about 20 cardboard toilet paper tubes (all approximately the same diameter)
- foam core, stiff cardboard, or mat board to serve as a base for the cardboard tubes, about 4 in × 30 in (10 cm × 75 cm)
- ruler
- poster board or mat board, approximately 8 in × 10 in (20 cm × 25 cm)
- pencil or pen
- pushpin
- scissors
- drinking straw
- 2 bamboo skewers
- paper clip
- string, about 12 in (30 cm)

1 Use hot glue to attach a cardboard tube at one end of the base. The length of the tube should be placed across the base, as shown in figure 1.

Figure 1

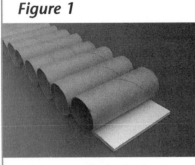

Toilet paper rolls glued to base

2 Continue gluing tubes to the base, with each tube just touching the one before it, until you reach the other end of the base.

3 Measure the diameter of three or four of the cardboard tubes. The diameters should be approximately $1\frac{11}{16}$ inches (4.3 cm). If this is the case, cut four square wheels from the poster board, with sides of 2 inches (5.1 cm). If the tubes you obtain have a significantly different diameter, then make the sides of the square wheels equal to 1.2 times the diameter (see Box o' Math at the end of this snack).

4 Locate the center of each square wheel by drawing two diagonals, as shown in figure 2.

Figure 2

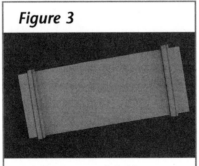

Using diagonals to locate the center of the square wheel

5 Poke a small hole in the center of each square wheel with a pushpin, taking care not to bend or crease the wheel.

6 From the poster board, cut out a 2- × 5-inch (5- × 12-cm) piece for the cart body.

7 Cut two sections of straw, each 2 inches (5 cm) long.

8 Hot glue the straw sections to the rectangular piece of poster board $\frac{3}{8}$ inch (1 cm) from each end, as shown in figure 3. This assembly will be the body of a small cart.

Figure 3

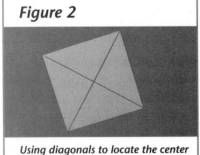

Drinking-straw axle holders

9 Cut the skewers into two 5-inch-long (12-cm) pieces, each with a point at one end. These will be axles. (If you are unable to cut the skewers with the scissors, just break the skewers or cut them with a utility knife.)

10 Slide one of the square wheels onto a skewer until it is about $\frac{3}{4}$ inch (2 cm) from the non-pointed end. Slide the pointed end through the straw, and then slide the other square wheel onto the skewer. Adjust the positions of the wheels so that they are aligned with each other and are fairly close to the edge of the cart. The wheel-and-axle assembly should turn freely in the straws. Assemble the other set of wheels the same way. When all the wheels are on, the cart should look like the one in figure 4.

Figure 4

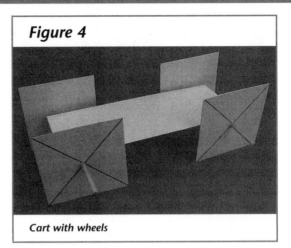

Cart with wheels

11 Use the pushpin to poke a hole in the body of the cart between the straw and one end, equidistant from the edges. Put one end of the paper clip through the hole, and adjust until it is positioned as shown in figure 5.

12 Tie a loop in the end of the string, and place it on the paper clip as shown in figure 5.

Figure 5

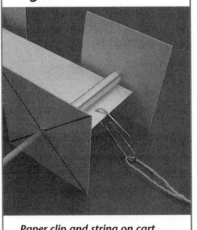

Paper clip and string on cart

Figure 6 Square wheels on the rounded road

To Do and Notice

Place the cart at one end of the cardboard-tube "road" and pull gently on the string so that the cart travels along the road. Notice that the cart rolls along smoothly and that the axles stay at a reasonably constant height.

What's Going On?

The cart rolls smoothly along the bumpy road because the vertical distance from each axle to the horizontal base of the road is always about the same. Each axle moves from a point above a low spot between two tubes (see figure 7) to a point above a high spot on a tube (see figure 8). The increasing height of the point on the circular tube where the tube contacts the wheel is compensated for by the decreasing distance on the wheel between the axle and the edge of the square where it contacts the tube. The same thing happens in reverse as each axle moves from a position above a high spot to a position above a low spot.

A special shape called a *catenary curve* (see Did You Know?), not a circle, is the curve that will give an absolutely level ride with square wheels. A road made with circles is a reasonably close approximation, however, and is easier to build from commonly available materials.

On a flat road, this would be the ultimate exercise bike! Mathematician Stan Wagon was inspired to build his square-wheeled tricycle after seeing the Exploratorium's Square Wheels exhibit.

So What?

A key problem in designing automobile transmissions involves gear teeth. Gear teeth must mesh together without slipping, because slipping results in frictional wear. In order for gears to mesh smoothly, engineers must design teeth that have matching shapes—a problem that's quite similar to designing the particular bumpy road that will provide a smooth ride for square wheels.

Figure 7

The height of the axle always remains about the same distance from the base (6.8 cm in figure 7 and 6.9 cm in figure 8). In figure 7, the vertical distance between the axle and the bottom of the wheel is maximum; because the wheel is in a depression, however, the remainder of the distance to the base is minimum. In figure 8, the situation is reversed. The vertical distance between the axle and the bottom of the wheel is minimum, but the remainder of the distance to the base is maximum.

Figure 8

Box o' Math
Calculating Wheel Size

To travel smoothly over the array of tubes, the sides of the square wheels have to be 1.2 times the diameter of the tubes. The equations below explain how this relationship is derived; the diagram shows you how the math applies to the square wheels and the "road." Note that l is the side of the square and d is the diameter of the circle (which represents the tube). The circumference of the tube $= 2\pi r$.

$$\cos 45 = \frac{AC}{AB}, \text{ or } AB = \frac{AC}{\cos 45} = \frac{r}{\cos 45}$$

$$AD = r$$

$$DB = AB - AD = \frac{r}{\cos 45} - r = r\left(\frac{1}{\cos 45} - 1\right) = r\left(\frac{1}{.71} - 1\right) = r(1.41 - 1) = 0.41r$$

$$l = \widehat{DF} + 2DE$$

$$\widehat{DF} = \frac{2\pi r}{4}$$

$$DE = DB = 0.41r$$

$$l = \frac{2\pi r}{4} + 2 \times 0.41r = 0.5 \times 3.14r + 0.82r = 1.57r + 0.82r = 2.4r$$

$$r = \frac{d}{2}, \therefore l = 2.4 \times \frac{d}{2} = 1.2d$$

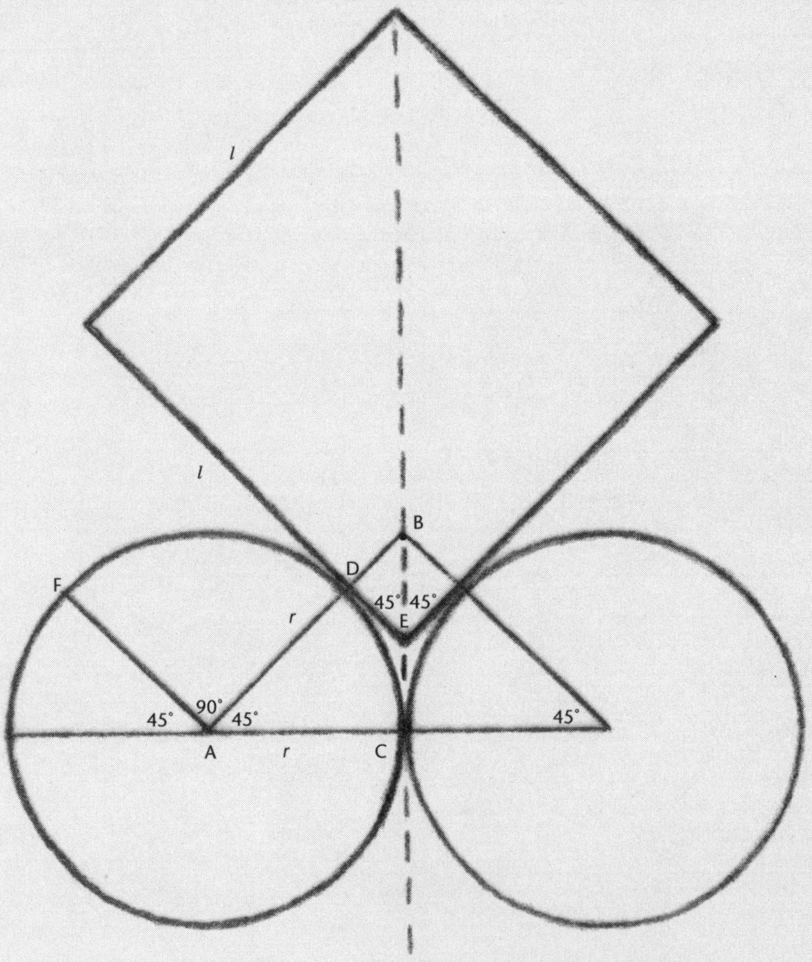

Square Wheels • An Exploratorium Science Snackbook

Figure 9 Two catenary curves

Did You Know?

Cat and Who?

A catenary curve is the shape a flexible rope or chain assumes when it hangs loosely and freely between two supports. Turned upside down, a catenary curve is the shape that will provide the greatest strength to an arch supporting only its own weight, such as the Gateway Arch in St. Louis.

Going Further

Deluxe Version

The article listed in the Credits & References section contains a template for a catenary curve and instructions for building a catenary road and a set of matching square wheels from plywood. If you have access to power tools, you might consider building this project.

Credits & References

This snack is based on the Exploratorium exhibit of the same name.

Regester, Jeffrey. "A Long and Bumpy Road." *The Physics Teacher,* April 1997. (Also reprinted in *Apparatus for Teaching Physics: A Collection of "Apparatus for Teaching Physics" Columns from The Physics Teacher,* 1987–1998, edited by Karl Mamola, American Association of Physics Teachers, 1998, pages 46–47.)

String Machine

Waves you never node about.

Waves are everywhere. They break on the shores of the ocean, bring music to your ears, and carry the signal of your favorite radio station. This snack—essentially a string attached to two small electric motors rotating in the same direction—allows you to create and play with a special class of waves called standing waves.

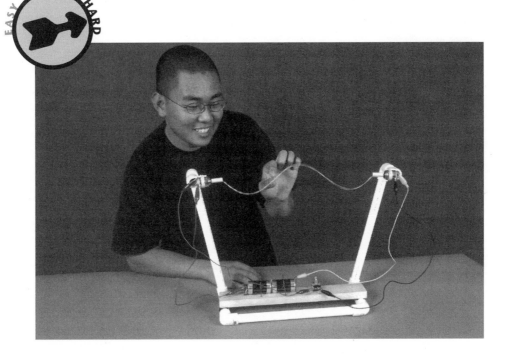

Materials

- $\frac{1}{2}$-in PVC pipe, 1 piece, 4 ft (1.2 m) long
- PVC shears (or hacksaw)
- electric band saw (or hacksaw)
- 8 each, $\frac{1}{2}$-in PVC 90° elbows
- hammer
- 6 finishing nails, $1\frac{1}{2}$ in long
- $\frac{3}{4}$-in standard pine shelving, $2\frac{1}{2}$ in × 15 in (6 cm × 38 cm); a 15-in-long piece of "1-×-3" board meets these dimensions
- 2 D-cell batteries
- 1 brass cup hook (or 1 additional nail)

- Velcro, 5 in (13 cm) long
- 25-ohm potentiometer (e.g., RadioShack #271-265); the potentiometer provides continuously adjustable speed control; if you want to omit it, you can still have two speeds—see Helpful Hint
- drill
- $\frac{1}{16}$-in drill bit
- $\frac{3}{8}$-in diameter wooden dowel, 2 segments, each $\frac{3}{4}$ in (2 cm) long
- 2 motors, 1.5–3 volts (e.g., RadioShack #273-223)
- 2 rubber bands

- piece of string, 18 in (46 cm) long (Braided string works significantly better than ordinary twisted string because it won't unravel. Wellington brand Braided Nylon Chalk and Mason Line works well and is available at some hardware stores; if you use regular twisted string, you will have to manually "retwist" it if it unravels.)
- masking tape
- 4 alligator-clip leads, 2 ft (60 cm) long (e.g., RadioShack #278-1157)

1 From the half-inch PVC pipe, cut two 2-inch-long pieces, two 4-inch-long pieces, and three 12-inch-long pieces (two 5-cm-long pieces, two 10-cm-long pieces, and three 30-cm-long pieces). Use PVC shears, a hacksaw, or an electric chop or band saw (PVC shears are an amazingly handy tool for cutting PVC pipe).

2 Use a hacksaw or band saw to slice the top portion off two PVC elbows, as shown in figure 1. These modified elbows will serve as cradles for the motors.

Figure 2

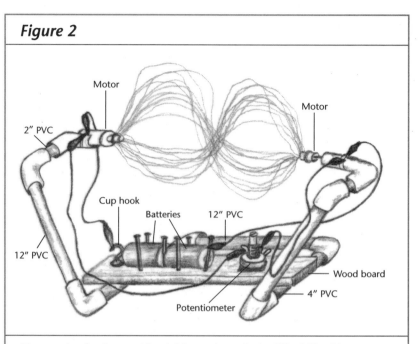

The completed string machine (with rotating string) will look like this.

Figure 1

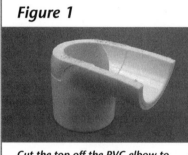

Cut the top off the PVC elbow to make the motor mount.

3 Using the PVC pipe pieces and elbows, create the PVC part of the structure shown in figure 2. Other parts will be added in succeeding steps. **NOTE:** Do not use glue to hold the PVC pieces together! You will need to adjust the joints when the String Machine is running, and the lack of glue allows the machine to be taken apart for storage.

4 Hammer the six nails into the board as shown in figure 3 to form a holder for the two D-cell batteries. One of the nails goes between the two batteries.

5 Screw the cup hook into the board so that when it's in the position shown in figure 3, it contacts the flat end of the battery and holds

Figure 3

Notice the nail placed between the two batteries in the battery holder.

it tightly in place. If you don't have a cup hook, you can use a seventh nail instead, but a nail sometimes doesn't make good contact with the flat end of the battery.

6 Stick 2-inch (5-cm) Velcro strips to the ends of the board and to the short PVC pieces in the base so that the board can be attached to the base.

7 If you are using the potentiometer for speed control, stick 1-inch (2.5-cm) pieces of Velcro to it and to

➡ Helpful Hint

Simple Speed Control

If you want to omit the potentiometer, you can get two different speeds by attaching the alligator clip (see step 13) to the end nail of the battery holder (two-battery speed) or to the middle nail (one-battery speed).

the board to hold the potentiometer in place on the board as shown in figure 2.

8 Use a $\frac{1}{16}$-inch drill bit to drill a hole $\frac{3}{8}$ inch (1 cm) deep in the center of the end of each piece of dowel. (The $\frac{1}{16}$-inch drilled hole is slightly smaller than the diameter of the motor shaft.)

9 Hold one of the motors so that the short end of the shaft (which protrudes from the plastic end-cap) is touching a hard surface and the long shaft on the other end is sticking straight up from the motor. Press the dowel down onto the motor shaft so that the shaft fits into the hole—it's a tight fit, but you want it to be tight so it won't slip. **NOTE: Do not try to push the motor shaft down into the dowel—you are likely to pop the motor right out of its casing.** Repeat this step for the other motor and other dowel piece.

10 Use rubber bands to hold the motors in place in the cradles you made when you sliced off the tops of the two elbows (see figure 4).

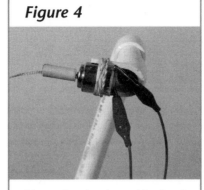

Figure 4

Motor, dowel, string, rubber bands, and alligator clip leads

11 Attach the ends of the string to the wooden dowels by laying each end of the string along a dowel and then wrapping masking tape around both (see figure 4).

12 Adjust the spread between the ends of the motor arms so that the string hangs in a loose curve.

13 Connect the alligator-clip leads as shown in figure 2, leaving one clip unattached. If you are using the potentiometer, connect one clip to its middle contact and the other clip to either one of the outer contacts. If you aren't using the potentiometer, just connect the two clips together.

14 Briefly touch the remaining unattached clip to its contact point to see if both motors are turning in the same direction. If they aren't, reverse the connections on one of the motors. (If neither motor turns, try turning the potentiometer knob. If that doesn't get them turning when you touch the clip to the contact point, check all the electrical connections carefully and make sure your batteries are working.)

15 Make the final connection, and adjust the potentiometer (if present) to obtain an intermediate speed.

To Do and Notice

Adjust the tension in the string (by adjusting the spread between the ends of the motor arms) and the motor speed (by turning the potentiometer knob) until you obtain a relatively stable pattern in the string. See if you can get the pattern to change to a different stable pattern by gently "pinching" the string: Hold your thumb below the string pattern and your forefinger above it, and slowly compress the pattern without actually making your fingers touch. Play with the machine for at least a few minutes to see how many different behaviors you can produce in the string.

While the machine is running, gently pull the motor ends of the two arms apart to increase the tension in the string. At some point, the string should snap into its simplest behavior mode, in which it looks something like a high-speed jump rope. This behavior is shown in figure 5. (Once you have this pattern, moving the arms back together a little usually helps to stabilize it; you can try further adjusting string tension or motor speed to get the best and most stable pattern.)

Figure 5 *The simplest wave pattern has two nodes.*

Notice that the string moves very little near the ends, but quite a lot in the middle. In wave language, a place in a wave with little or no movement is called a *node,* and a place with maximum movement is called an *antinode.*

Gently pinch the string near the middle—or just press down on it with a pencil. With a little practice (and perhaps some adjustment of string tension or motor speed), you should be able to make the string jump into a mode with three nodes (one at each end, and one in the middle), and two antinodes (one in the middle of each loop) as shown in figure 6.

Figure 6 *This wave pattern has three nodes and two antinodes.*

Double-Dutch rope jumpers make patterns like those made by your string machine.

time, the two waves traveling in opposite directions combine with each other, adding together to produce a single overall shape for the rope. This is still true even if both you and your friend shake the rope and each of you produce wave pulses. If you both shake the right way, you can produce an overall shape that is stable and exhibits nodes and antinodes. This stable overall shape is called a *standing wave.* The various stable patterns you produced with your String Machine are also standing waves.

The simplest standing wave that you produced was the fundamental, or first harmonic, as shown in figure 5. The fundamental is one-half of a whole wave, or one-half a wavelength (see figure 8). The second standing wave you made, the one with three nodes, is a second harmonic wave, and it is a whole wavelength. The third wave you made, with four nodes, is a third harmonic wave, which is one-and-a-half wavelengths.

For waves in a string, standing wave formation normally depends on a number of factors, including the frequency with which the string is shaken and the tension in the string. If any of these factors is changed, then the standing wave pattern changes. In your String Machine, however, it's possible to change the frequency (motor speed) and tension (spread between ends of the motor arms) *without* changing the pattern because of the circular motion of the string. This motion introduces forces on the string not present in waves generated in a single plane, as in the simple case of the jump rope being shaken up and down.

What's Going On?

As the dowel turns on the motor shaft, the end of the string that is taped to the edge of the dowel moves in a circle. If you think of circular motion as a combination of vertical and horizontal motion, you can visualize the string as being shaken up and down at the same time as it is being shaken right and left. The shaking of the string causes wave pulses to travel along the string.

You and a friend can produce wave pulses in a jump rope that are very similar to those in the string. If the person holding the other end of the rope holds the rope tightly while you shake the rope, the pulses you make will bounce off the person's hand and travel back toward you. These returning pulses will travel through the ones you continue to make. Since any particular piece of the rope can only be in one place at one

Now try pinching the string about one-third of the way across. With a little practice, you should be able to make the string jump into a more complicated pattern with three loops. This pattern, shown in figure 7, has four nodes and three antinodes.

Figure 7 *Try making a more complicated wave pattern with four nodes and three antinodes.*

Put the string back into the mode with three nodes and two antinodes as shown in figure 6. Spread the fingers on one of your hands slightly apart so you can see between them, and wave this hand back and forth between the string and your eyes so that you can see the string between your fingers. Can you make the string seem to stand still? If not, reduce the motor speed and try again. (Closing one eye may also help.) Eventually you may be able to see a single wave, rather than the blurred pattern.

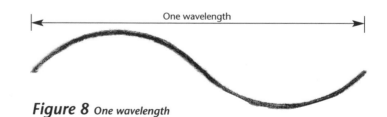

One wavelength

Figure 8 *One wavelength*

Square Wheels • An Exploratorium Science Snackbook

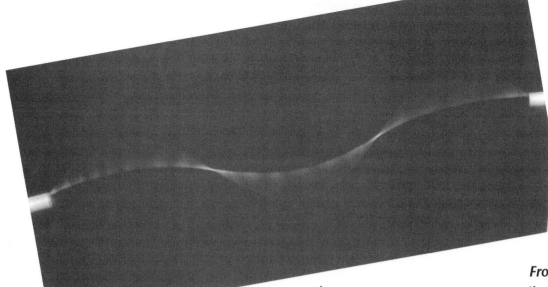

Figure 9 *Try a digital camera for instant results!*

When you "strobed" the string by waving your hand back and forth in front of it, you were able to get successive views of the string over short time intervals. If the time it took for adjacent spaces between your fingers to change places in front of your eyes was exactly the time it took for the string to go through one whole cycle (or any whole number of its cycles), then you saw the wave in the same position each time, and it appeared to be standing still.

So What?

Standing waves are at the heart of musical instruments. In wind instruments, standing waves are set up in an air column. At some locations, the density of air molecules is alternately very high and very low, creating large fluctuations in pressure. These locations are the antinodes of the air-pressure standing wave. At other locations, small pressure changes mark the nodes of the standing wave. When the length of the air column is changed, as with keys on a clarinet or the slide on a trombone, a different standing wave is formed, and you hear a different note. Standing waves are also formed on stringed instruments and drumheads.

Did You Know?

Getting in Tune

On occasion—possibly due to slightly different motor speeds—the moving string may seem to develop sub-patterns within the main pattern, which will vary slowly in a regular way. This behavior is the result of waves with slightly different frequencies interacting in a complex way to produce a regular alternating or oscillating pattern called a *beat*. Perhaps the simplest and most well-known example of a beat is the loud-soft-loud-soft tone produced when a musical instrument (or two different strings on the same instrument) are being tuned. The two instruments are in tune—each producing the same frequency note—when the variable tone is no longer noticeable.

Harmonics and Overtones

The second harmonic is also called the first overtone, the third harmonic the second overtone, and so on.

Going Further

Getting Loopy

What is the maximum number of loops you can produce in the string? You may try maintaining a continuous loose pinch or exerting continuous light pressure with your finger or a pencil.

Frozen Waves

If you have a strobe light, try strobing the string on your String Machine—it's spectacular! A good camera can produce images of a "frozen" string as well, as shown in the opening photo at the beginning of this snack and in figure 9. Using a digital camera will provide an instant picture of the string in mid-wave.

Rainbow Waves

Build a motorized color wheel, with transparent pie-shaped segments of a few different colors. If you shine a bright light through the rotating color wheel and use it to illuminate the string machine while the string is in motion, the results can be sensational.

Credits

The String Machine is a low-budget, black-and-white version of the commercially available, full-color String Ray.

Don Rathjen, Tien Huynh-Dinh, and Guillermo Trejo-Mejia contributed to the design of this snack.

Stripped-Down Generator

If you shake just right, you'll see the light.

In an electric power plant, steam or water power is used to move huge coils of wire past extremely strong electromagnets, generating megawatts of electricity to light whole towns. In this snack, you use your muscles to move ordinary magnets past a small coil of wire, generating milliwatts of electricity—just enough to light an LED. Although the two generators work at very different scales, they're based on the same physics principles.

Materials

- 2 blank overhead transparency sheets, $8\frac{1}{2}$ in × 11 in
- 2 film canisters with lids
- tape (transparent or translucent)
- 3 rubber bands, $\frac{1}{8}$-in (3-mm) wide × $3\frac{1}{2}$-in (9-cm) long (#32 or #33)
- about 200 ft (60 m) of #30 magnet wire (very thin copper wire with enamel insulation); RadioShack sells magnet wire in packages containing one spool

each of #22, #26, and #30 wire; the #30 spool has 200 ft of wire on it (RadioShack #278-1345)
- scissors, knife, or wire cutters
- small piece of sandpaper
- 2 mini alligator clips (RadioShack #270-380A, pack of 12 clips)
- light-emitting diode (LED); some LEDs give particularly bright light (e.g., RadioShack Jumbo Super-Bright LED, #276-086, and High Brightness Red LED, #276-066);

many ordinary LEDs will also work; RadioShack sells a variety of LEDs, including a 20-pack of assorted LEDs, #276-1622
- 5 ceramic "donut" disk magnets, $1\frac{1}{2}$ in diameter (RadioShack #54-1888, pack of five)
- bipolar (two-color) LED (RadioShack #276-012)
- masking tape

1 Put the two transparency sheets together, one on top of the other, and roll them into a tube, with the long side of the sheets running the length of the tube. Place a film canister (with lid on) in each end of the tube, with the top of the canister facing out. Tighten the tube so that it fits snugly against the canisters at both ends, and tape it at the ends and in the middle so that it stays this diameter. Also tape along the lengthwise seam for added strength. You should now be able to remove the canisters and reinsert them without any trouble.

2 Double two of the rubber bands and place them on the tube so that they are wrapped around the tube near the middle, with about 1 inch (2.5 cm) between them. The rubber bands will help keep the coils of wire you will be wrapping around the tube from spreading.

3 Wrap the magnet wire around the tube between the rubber bands, leaving about 12 inches (30 cm) of wire free on the starting end. Begin as close as you can to one rubber band and proceed toward the other rubber band. When you reach the second rubber band, start another layer and proceed back toward the original rubber band—but keep wrapping in the same direction (i.e., clockwise or counterclockwise, whichever direction you began with).

If you reverse the direction of your wraps, you'll cancel the effect of the wire you wrapped initially.

Keep wrapping the wire around the tube, building up multiple layers, until you have about 500 wraps. (If you use the #30 magnet wire from RadioShack, you'll get about 450 to 500 wraps if you use the whole spool—anything close to this will be enough.) Wrap the wire as tightly as you can without squashing the tube. Be sure you don't scrape the insulation off the wire as you wrap it. Try to keep all the wire between the rubber bands. (If the coil starts to come apart, use tape to hold it in place.) Make sure to leave another 12 inches (30 cm) of wire free.

4 You should now have a coil with two 12-inch pieces of wire coming from it. Tape the base of each free wire end to the tube near the coil to hold it in place. Then cut both of these wires so that they extend about 3 inches (8 cm) from the coil.

5 Use a knife or sandpaper to remove about $1\frac{1}{2}$ inches (4 cm) of the insulation from each of the two ends of wire. Make sure that you remove the insulation thoroughly. Put the stripped end of one of the wires through the hole in the shank of an alligator clip, and then wind the wire tightly around the shank of the clip. Be sure that the stripped wire makes

good contact with the clip. Repeat for the other wire and alligator clip.

6 Clip each alligator clip to one of the legs of the LED. Lay the alligator clips alongside each other on the tube, but be sure that they don't touch each other. Use the third rubber band to hold them in place. The completed coil and LED assembly should look like figure 1.

Figure 1

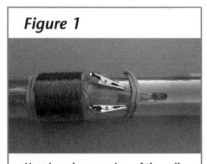

Here is a close-up view of the coil, alligator clips, and LED assembly.

7 Put the donut magnets together to form a stack. Remove one of the film canisters from the tube, put the magnets in the tube, and replace the canister in the end of the tube. (Be sure the lid is still on the canister, so that squeezing the canister won't easily deform its circular cross-section.) Tape the film canisters in place with masking tape. Figure 2 shows the completed Stripped-Down Generator.

Figure 2

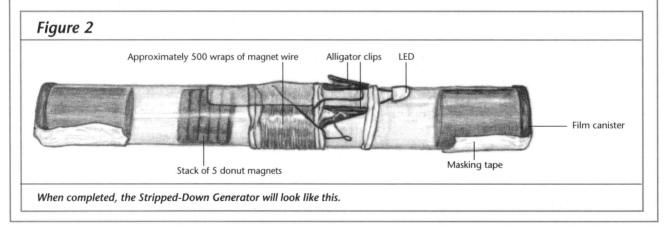

Approximately 500 wraps of magnet wire · Alligator clips · LED · Film canister · Stack of 5 donut magnets · Masking tape

When completed, the Stripped-Down Generator will look like this.

To Do and Notice

Grasp the tube at both ends to hold the film canisters in place. Shake the tube back and forth horizontally so that the magnet stack slides back and forth inside the tube. Shake the tube as rapidly as you can.

Each time the magnet stack passes through the coil of wire, the LED should flash. The flashes are more dramatic in a darkened room.

Remove the original LED and hook up the bipolar LED. Shake the tube again so that the magnet stack slides back and forth. What happens now?

What's Going On?

Whenever a wire and a magnetic field move perpendicular to each other, a voltage is induced in the wire. If the wire is part of a complete electric circuit, the voltage will cause a current to flow in the circuit. If a coil of wire is used instead of a single wire, the voltage obtained is the single-wire voltage multiplied by the number of turns in the coil.

In the Stripped-Down Generator, every time the magnet stack moves through the coil of wire, the coil experiences a changing magnetic field, which induces a voltage in the coil. Since the coil is part of a complete circuit that includes the LED, current flows in the LED and it lights. Note, however, that this depends on your shaking the generator rapidly. If you shake it too slowly, you may not produce the voltage needed to light the LED.

That's the simple explanation. What's actually going on is a little more complicated. The stack of magnets has two poles, and the lines of magnetic force are oriented differently at each pole. When the north pole of the stack passes through the coil, a voltage is created. Then when the south pole passes through, the voltage is reversed. Therefore, each time the magnet passes through the coil,

two pulses of voltage are produced, and they are opposite each other in sign.

The normal LED lights up once each time the magnet stack passes through the coil of wire, regardless of which way the stack is traveling, even though two pulses of voltage are produced. There is only one flash because the LED is a diode, and a diode only allows current to flow in one direction. When the magnet stack passes by the coil in one direction, the first pulse of current lights the LED, and when it passes by in the other direction, the second pulse of current does the job.

The bipolar LED is actually two LEDs in one casing. When current flows in one direction, one of the LEDs emits red light; when current flows in the opposite direction, the other LED emits green light. When you hook up this LED, you see two flashes of light each time the magnet passes through the coil—one of them when the north pole of the magnet stack passes through the coil, and the other when the south pole passes through.

More About the Two Pulses

When the magnet stack passes through the coil, two pulses of voltage of opposite sign are produced. This happens because the direction of the voltage induced in a coil is related to both the *orientation* of the magnetic field lines relative to the coil, and their *direction of movement* relative to the coil. When the north pole of the magnet stack, for example, is moving to the left through the coil, as shown in figure 3a, the magnetic field lines are

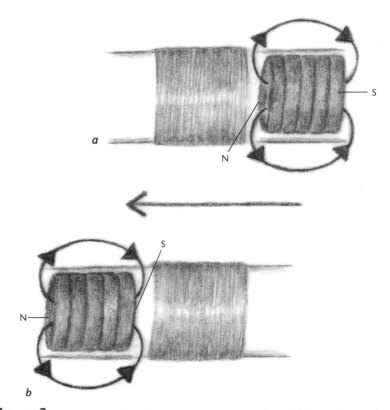

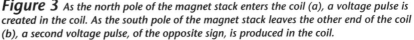

Figure 3 *As the north pole of the magnet stack enters the coil (a), a voltage pulse is created in the coil. As the south pole of the magnet stack leaves the other end of the coil (b), a second voltage pulse, of the opposite sign, is produced in the coil.*

The difference in sign of the voltage pulses is due to the different direction of the magnetic field lines in each case. At the north pole of the magnet stack, the magnetic field lines are pointing outward, and at the south pole they are pointing inward. When the LED is connected in the proper direction relative to the sign of the voltage pulse, current flows in the LED, causing it to light.

In this large electric generator, coils of wire spin around in a magnetic field, generating voltage and currents.

Did You Know?

The Motion Is Relative

When you shake the Stripped-Down Generator, the coil is moving back and forth over the magnet at least as much as the magnet is moving back and forth through the coil. In any case, it's the relative motion of coil and magnet that's important in generating electricity.

Going Further

Dual LEDs

Attach an ordinary LED to the alligator clips. Note which clip holds the longest leg of the LED. Then hook up a second ordinary LED, with its longest leg in the other clip. The two LEDs now have opposite polarities. Predict what will happen when you shake the tube.

Measure It

Hook the generator up to a voltmeter (not digital) and see what the needle does.

Picture It

Hook the generator up to an oscilloscope and see if you can display the entire signal.

pointing *away* from the center of the coil, and the resulting voltage forces the current to flow in one direction in the coil. Then when the south pole of the magnet stack passes through the coil, as shown in figure 3b, the magnetic field lines are pointing *toward* the center of the coil, and the resulting voltage forces the current to flow in the opposite direction.

So What?

Current that always flows in the same direction is called direct current (DC). Current that flows first in one direction and then the other is called alternating current (AC). The generators that pro-vide the electricity you use every day operate on the same general principle as the Stripped-Down Generator, and they produce AC current just like the Stripped-Down Generator.

Most of the appliances and lights in your home can use AC current straight from the outlets. Some electrical devices, however, require DC rather than AC for their operation, and they either have a built-in electrical circuit that converts the AC to DC, or they use an adapter that accomplishes this task. These devices include portable phones, answering machines, stereos, and computers. Many devices that can be run on batteries, such as laptop computers and tape recorders, also come equipped with AC adapters, which convert the AC from the wall outlet into DC for the device.

Credits & References

Curt Gabrielson developed this snack based on an idea proposed by a sixth grade student named Van in an afterschool workshop.

Macaulay, David. *The Way Things Work.* Boston: Houghton Mifflin, 1988. See LED, page 293, and Electric Generator, pages 304–305.

Your Father's Nose

Sometimes you are the splitting image of the person across the table.

In a normal mirror, you see your own face looking back at you. But what if you looked into a mirror and saw a face that was partly yours and partly your friend's? A mirror with horizontal gaps in it will allow you to have this odd and amusing experience.

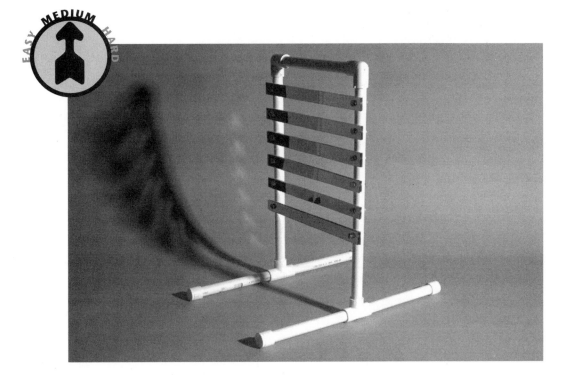

Materials

- PVC shears or hacksaw
- $\frac{1}{2}$-in PVC pipe, Schedule 40, 69 in (1.75 m) long
- ruler
- pencil
- electric drill or drill press
- $\frac{3}{16}$-in drill bit
- 12 plexi-mirror strips, 1 in × 12 in long (2.5 cm × 30 cm). Plexi-mirror is plastic mirror and is available at plastics stores, which you can locate through the yellow pages of your phone book; it can be cut with a band saw, with a plywood blade on a table saw, or you can get it cut at the store
- clamps to hold material in place when drilling (if you haven't had experience drilling PVC pipe or plexi-mirror, ask for advice or help)
- 2 PVC 90° elbows, $\frac{1}{2}$ in
- 2 PVC T-joints, $\frac{1}{2}$ in
- 4 PVC caps, $\frac{1}{2}$ in
- 12 brass fasteners (roundhead fasteners), $1\frac{1}{2}$ in long (see figures 1 and 2)
- a partner

Figure 1

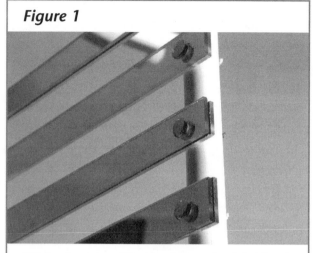

This is a front view of the brass fastener heads holding the plexi-mirror strips.

Figure 2

Here is a back view of the same brass fastener heads.

1 Using the PVC shears or hacksaw, cut the PVC pipe into two 18-inch pieces, one 9-inch piece, and four 6-inch pieces (two 46-cm pieces, one 23-cm piece, and four 15-cm pieces).

2 Draw a straight line along the entire length of each of the 18-inch (46-cm) pipes.

3 At a point on the line 2 inches (5 cm) from one end, drill a $\frac{3}{16}$-inch hole completely through both walls of one of the 18-inch (46-cm) pipes. Drill five more similar holes along the line, each 2 inches (5 cm) from the last one. Be sure the holes go straight through the pipe. Drill an identical set of six holes in the other 18-inch (46-cm) piece of pipe.

4 Assemble the PVC frame as shown in the opening photo. The two 18-inch (46-cm) pieces are the vertical legs, the 9-inch (23-cm) piece is the horizontal top, and the four 6-inch (15-cm) pieces form the horizontal feet of the base. Orient the holes in the 18-inch (46-cm) pipes so that a nail stuck through a pair of holes would be parallel to the feet of the frame.

5 Measure the distance across the frame between the highest holes on each pipe.

6 Place two of the plexi-mirror strips back to back, and drill holes through them the same distance apart as the distance between the highest holes on the frame.

7 Repeat step 6 for the other five pairs of mirror strips.

8 Use the brass fasteners to attach the back-to-back pairs of mirror strips to the PVC frame as shown in figures 1 and 2.

9 If there is any bowing of the mirror strips, try twisting the vertical PVC pipes to adjust the mirrors. If the bowing persists, try drilling slightly larger holes in the mirror strips that bowed.

To Do and Notice

Sit down at a small table with another person directly across from you and put the assembled frame on the table between you as shown in figure 3.

Figure 3 *Look through the mirror strips at a friend's face. You'll see one face that combines the features of your two faces.*

With your face about 12 inches (30 cm) from the mirror strips, move your head up or down until you can see your eyes in the second mirror strip from the top. Have the person on the other side hold his or her head the *same* distance away from the mirror strips on the other side. Then have the person move his or her head up or down until he or she can see his or her eyes in the other side of the same mirror strip that you are looking into.

You should now see a "composite" face made up of some of your features and some of the other person's. To get the best composite face, you may need to move closer to or farther from the mirror while your partner remains stationary. You may also have to move your head up or down or sideways. Experiment and see what happens.

Move your head downward until your eyes are looking through the space directly below the mirror strip you were looking in previously. Have the other person move his or her head the same way. How does this composite face compare with the previous one? Move your heads back and forth between the two positions to compare the two composite faces a few times.

Try it again with another person.

What's Going On?

When you see your eyes in the mirror, you may see the other person's mouth through the gap. At the same time, the other person will be seeing his or her own eyes, but your mouth. You see a "composite" face made up of alternating horizontal bands of your own features and the other person's features. The other person has a similar experience.

Your image in the mirror is the same distance "behind" the mirror as your actual face is in front of the mirror. To get your facial features to blend with those of the other person's, therefore, that person must be just as far away from the mirror as you are, so that his or her face is in the same place as the virtual image of your face.

In **The Blank Signature,** *artist René Magritte (1898–1967) investigates the idea of an image broken up into strips.*

So What?

It may seem odd that your reflection appears to be behind the surface of a plane (flat) mirror rather than at the surface of the mirror. To understand why that's so, take a look at figure 4.

Light bounces off the tip of the nose of a man looking into a mirror. That light reflects off the mirror and enters both the man's eyes. The man's eyes and brain work together to make a picture of the world. This eye-brain system assumes that the light has traveled straight lines to reach the eyes. The dotted lines in figure 4 show where the man's head would have to be for light to travel in a straight line to the eye. These lines place the image of the man's head behind the mirror.

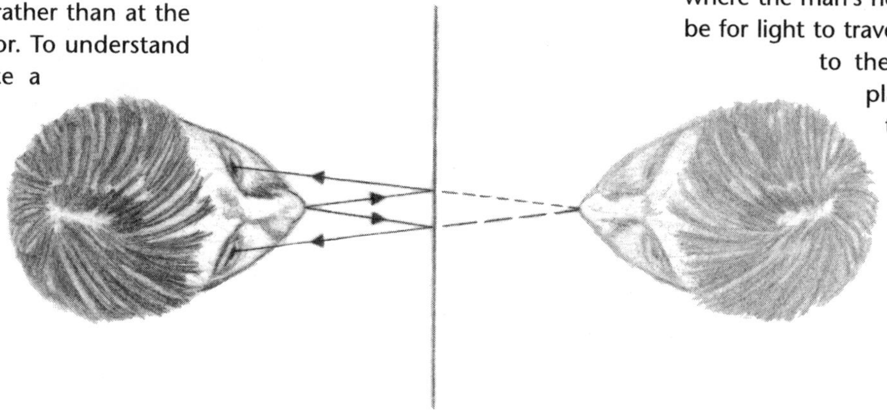

Figure 4 In a plane mirror, the image of a face is the same distance behind the mirror as the face is in front of the mirror.

Credits

This snack is based on the Exploratorium exhibit of the same name.

Resource Guide

In your hunt for primary resources, we strongly recommend that you browse through hardware or home improvement stores, office supply stores, discount stores, toy stores, thrift stores, science museum stores, even grocery stores, and think about how the items on display might be used in a science class. This can be an amazingly creative endeavor and can generate long-term rewards.

Thumbing through science catalogs is also an extremely worthwhile experience. You will get an overview of what things are available and what they are used for, and you may get ideas for how to improvise more economical substitutes or more creative variations. You can obtain catalogs by contacting the companies (or visit your local high school science department, where they often exist in abundance).

Beyond these general resources, the specific resources we have listed below are ones we have found to be useful over the years when undertaking the sorts of projects and activities that are featured in this book. We have listed the resources as either *Materials and Supplies* or *Books, Manuals, and Journals*. This collection is not intended to be all-encompassing, exhaustive, or universal, but we hope it will be useful.

Finally, we encourage you to consider the Exploratorium Web site as a resource, particularly the Web page devoted to snacks: www.exploratorium.edu/snacks. Here you will find more than 100 additional snacks, including most that appeared in the original *Exploratorium Science Snackbook.*

Materials and Supplies
(Sources are listed alphabetically)

American Science and Surplus
P.O. Box 1030, Skokie, Ill. 60076
847-982-0870
www.sciplus.com
Useful science stuff (lab equipment, small motors, etc.), hardware odds and ends, wide range of strange items for creative use.

Arbor Scientific
www.arborsci.com
800-367-6695
P.O. Box 2750, Ann Arbor, Mich. 48106
Nice selection of interesting science materials.

Dowling Miner Magnetics Corp.
P.O. Box 1829, Sonoma, Calif. 95476
707-935-0352 or 800-MAGNET 1
Neodymium magnets at reasonable cost. Cow magnets are also available.

Edmund Scientific Company
101 E. Gloucester Pike, Barrington, N.J. 08007
800-728-6999
www.edsci.com
A prime source for interesting science materials, from moiré patterns to solar cells to polarizing material. They are a particularly good source for optics supplies.

Educational Innovations
362 Main Ave., Norwalk, Conn. 06851
203-229-0740
www.teachersource.com
Nice selection of interesting science materials.

The Exploratorium Store
3601 Lyon St.
San Francisco, Calif. 94123
www.exploratoriumstore.com
Also check science museum stores in your own area.

Flinn Scientific
131 Flinn St., P.O. Box 219, Batavia, Ill. 60510-0219
708-879-6900
www.flinnsci.com
In addition to being a chemical supply house, Flinn has an unmatched array of materials related to laboratory safety. Their catalog is a valuable teaching resource, and they also offer publications with ideas for chemistry lab experiments and activities. Don't fail to visit their booth if you ever go to a National Science Teachers Association convention.

Frey Scientific
Beckley Cardy Group
100 Paragon Parkway, Mansfield, Ohio 44903
888-222-1332
www.beckleycardy.com
Large distributor of science supplies.

Kelvin
280 Adams Blvd., Farmingdale, N.Y. 11735
800-535-8469
www.kelvin.com
Outstanding source for technology, electronics, project, and science materials—from electric motors to hydraulic syringes to potentiometers to plastic propellers.

Local Feed Stores
Cow magnets are strong, cylindrical magnets with rounded ends. These magnets are fed to cows so that any iron material eaten by the cow (nails, bits of wire, and so on) will remain in the stomach and not pass through the digestive tract. These are excellent magnets for many science activities; you may be able to find them at a local animal feed store.

The Magnet Source/Master Magnetics
607 S. Gilbert, Castle Rock, Colo. 80104
800-525-3536
www.magnetsource.com
General source for magnets, including neodymiums.

Mouser Electronics
958 N. Main, Mansfield, Tex. 76063
800-346-6873
Mail-order electronics supplier with an extensive selection. Catalog sometimes routinely sent to schools or available upon request.

Plastics Stores
You can get plexi-mirror, clear and colored acrylic sheets (for static electricity demonstrations or colored filters), aluminized Mylar, and plastic tubes at plastics stores. Most will cut plastic to size (for a price), and they may have a bargain scrap bin for pieces by the pound. Check the yellow pages under Plastics.

RadioShack
Stores throughout the country. Catalog available.
800-THE-SHACK
www.radioshack.com
Since it is virtually everywhere, RadioShack is a unique source of common electronic components such as resistors, alligator clips, buzzers, wire, LEDs, etc. Because of its easy accessibility, we have given RadioShack part numbers for the electronic parts used in this book, but electronic components can also be obtained by mail (sometimes at lower prices) from catalog companies (e.g., see Kelvin and Mouser listings) and from local or regional electronics stores.

Sargent-Welch
P.O. Box 5229, Buffalo Grove, Ill. 60089
800-727-4368
www.sargentwelch.com
Large distributor of science supplies.

The Science Source
P.O. Box 727, Waldoboro, Maine 04572
800-299-5469
www.thesciencesource.com
Design, technology, and science products.

Books, Manuals, and Journals
(The following are listed alphabetically by author.)

The Physics Teacher
American Association of Physics Teachers (AAPT)
One Physics Ellipse, College Park, Md. 20740
301-209-3300
www.aapt.org
An outstanding magazine for physics teachers. Catalog of publications and materials available.

ChemMatters
American Chemical Society
Dept L-0011, Columbus, Ohio 43268
www.acs.org
An outstanding and very inexpensive quarterly magazine for chemistry teachers and students. Catalog of publications and materials available.

Physics Olympics Handbook
By Susan Arguso, Carole Escobar, and Virginia Moore
American Association of Physics Teachers (AAPT), 1984
Activities and procedures for holding a Physics Olympics.
American Association of Physics Teachers (AAPT)
One Physics Ellipse, College Park, Md. 20740
301-209-3300
www.aapt.org

How Things Work: The Physics of Everyday Life (2d ed.)
By Louis Bloomfield
New York: John Wiley & Sons, 2001
The content of physics approached through commonly encountered applications and devices, including light bulbs, microwave ovens, CD players, roller coasters, vacuum cleaners, cars, surfing, airplanes, and many more. A unique and outstanding addition to your reference library.

Clouds in a Glass of Beer: Simple Experiments in Atmospheric Physics, 1987
What Light Through Yonder Window Breaks?: More Experiments in Atmospheric Physics, 1991
By Craig F. Bohren
New York: John Wiley & Sons
www.wiley.com
Two books of discussion and activities concerning atmospheric physics.

Physics: A Window on Our World (4th ed.), 2001
The Physics Around You, 2002
By Jay Bolemon
New York: McGraw Hill
Two excellent additions to your reference library.

Explorabook
By John Cassidy and The Exploratorium
Palo Alto, Calif.: Klutz Press, 1991
Tools and ideas to inspire scientific exploration, written in the inimitable Klutz style.

A Potpourri of Physics Teaching Ideas: Selected Reprints from "The Physics Teacher," April 1963–December 1986
Donna Berry Conner, editor
Large collection of outstanding experiments, activities, and demonstrations from *The Physics Teacher* magazine.
American Association of Physics Teachers (AAPT)
One Physics Ellipse, College Park, Md. 20740
301-209-3300
www.aapt.org

How Things Work
By Richard Crane
American Association of Physics Teachers (AAPT), 1992
Collection of explanations of how many interesting things work, from the author's column in *The Physics Teacher.*
American Association of Physics Teachers (AAPT)
One Physics Ellipse, College Park, Md. 20740
301-209-3300
www.aapt.org

Physics (5th ed.)
By John Cutnell and Kenneth Johnson
New York: John Wiley & Sons, 2001
Excellent addition to a reference library.

String and Sticky Tape Experiments
By Ronald Edge
American Association of Physics Teachers (AAPT), 1987
Extensive collection of simple experiments, activities, and demonstrations from the author's column in *The Physics Teacher* magazine.
American Association of Physics Teachers (AAPT)
One Physics Ellipse, College Park, Md. 20740
301-209-3300
www.aapt.org

Turning the World Inside Out & 174 Other Simple Physics Demonstrations, 1990
Why Toast Falls Jelly-Side Down: Zen and the Art of Physics Demonstrations, 1997
By Robert Ehrlich
Princeton, N.J.: Princeton University Press
Two collections of physics experiments, demonstrations, and activities.

Thinking Physics (2d ed.)
By Lewis Carroll Epstein
San Francisco: Insight Press, 1989
Illustrated, multiple-choice conceptual physics problems related to the real world (includes solutions).

Exploratorium Magazine (formerly *Exploring* magazine)
The Exploratorium
3601 Lyon St., San Francisco, Calif. 94123
www.exploratorium.edu/exploring
This magazine is free as part of an Exploratorium membership. Each issue deals with a single topic, some of which have been Amusement Parks, Electricity, Illusions, Dirt, and Ice. While this magazine is not specifically focused on the classroom, the articles contain lots of unusual and useful information, and most issues have at least one "To Do & Notice" activity related to the topic of that issue.

Seeing the Light
By David Falk, Dieter Brill, and David Stork
New York: John Wiley & Sons, 1986
www.wiley.com
A unique and outstanding text covering virtually all aspects of light, vision, and color. Though essentially a college text, it is largely qualitative rather than mathematical. It also has many application activities that are quite doable by students. For any science teacher teaching the topic of light.

A Demonstration Handbook for Physics
By G. D. Frier and F. J. Anderson, 1981
American Association of Physics Teachers (AAPT), 1981
Brief descriptions and diagrams for zillions of physics demonstrations.
American Association of Physics Teachers (AAPT)
One Physics Ellipse, College Park, Md. 20740
301-209-3300
www.aapt.org

Physics (5th ed.)
By Douglas Giancoli
Englewood Cliffs, N.J.: Prentice Hall, 1998
Fine addition to your reference library.

Conceptual Physical Science (2d ed.)
By Paul Hewitt, John Suchocki, and Leslie Hewitt
Menlo Park, Calif.: Addison Wesley Longman, 1999
www.awl.com
Outstanding conceptual treatment of physical science.

Conceptual Physics (9th ed.)
By Paul Hewitt
San Francisco, Calif.: Addison Wesley, 2002
www.awl.com
An outstanding example of a readable, qualitative approach to the concepts of physics. A classic that is a must for your reference library.

Conceptual Physics: A High School Physics Program (3d ed.)
By Paul Hewitt
Menlo Park, Calif.: Addison Wesley, 1999
www.awl.com
A Teachers Guide, Lab Manual, and other related teaching materials are available.

GEMS (Great Explorations in Math and Science)
Lawrence Hall of Science
University of California Berkeley, Berkeley, Calif. 94720
www.lhs.berkeley.edu//publications
A series of activity-based publications covering everything
from bubbles to fingerprinting.

Invitations to Scientific Inquiry (2d ed.)
By Tik L. Liem
Science Inquiry Enterprises, 1981
An outstanding collection of simple demonstrations and
activities; a high-priority acquisition. Don't leave a
methods course without it!
Science Inquiry Enterprises
14358 Village View Lane, Chino Hills, Calif. 91709
909-590-4618
jmkris@scienceinquiry.com

The Way Things Work
By David Macaulay
Boston: Houghton Mifflin, 1988
Profusely illustrated descriptions of the way a lot of
things work.

The Role of Toys in Teaching Physics
By Jodi and Roy McCullough
American Association of Physics Teachers (AAPT), 2000
An AAPT/PTRA Workshop Manual. Ideas for using a wide
variety of toys in labs, demonstrations, displays, contests,
and more. Includes sources for toys, concepts covered,
questions, and other useful items.
American Association of Physics Teachers (AAPT)
One Physics Ellipse, College Park, Md. 20740
301-209-3300
www.aapt.org

Getting Started in Electronics
By Forest M. Mims III
RadioShack, 1983
www.radioshack.com
A good introduction to basic electronics. Includes theory
and practical applications, and lots of stuff to build with
relatively inexpensive parts.

The Dick and Rae Physics Demo Notebook
By Richard B. Minnix and D. Rae Carpenter, Jr.
Lexington, Va.: Dick and Rae, Inc., 1993
VMI Mallory Hall, Lexington, Va. 24450
Physics demonstrations from two decades of workshops at
Virginia Military Institute.

The Science Teacher magazine
Science Scope magazine
Science and Children magazine
National Science Teachers Association (NSTA)
1742 Connecticut Ave., NW, Washington, D.C. 20009
www.nsta.org
Useful magazines designed for high school, middle school,
and elementary school science teachers, respectively.

NSTA Recommends catalog
Annual NSTA Guide to Science Education Suppliers catalog
National Science Teachers Association (NSTA)
1742 Connecticut Ave., NW, Washington, D.C. 20009
www.nsta.org

Chemical Magic from the Grocery Store
By Andy S.W. Sae
Dubuque, Iowa: Kendall/Hunt, 1996
An excellent collection of sixty activities and
demonstrations using common, easily obtainable
materials.

Teaching Chemistry with Toys: Activities for Grades K–9
By Jerry L. Sarquis, Mickey Sarquis, and John P. Williams
New York: McGraw-Hill Trade, 1995
Another excellent collection of activities based on toys.

Physics from the Junk Drawer & Counter Top Chemistry
Science House Staff, North Carolina State University
Dubuque, Iowa: Kendall/Hunt, 1997
Nice collection of hands-on physical science activities for
middle school students, teachers, and others interested in
science.

Teaching Physics with Toys: Activities for Grades K–9
By Beverley Taylor, Jim Poth, and Dwight Portman
New York: McGraw-Hill Trade, 1995
An excellent collection of activities based on toys.

TOPS Task Cards
TOPS Learning Systems
10970 Mulino Rd., Canby, Ore. 97013
www.topscience.org
Activity cards for physical science projects using everyday
materials (rubber bands, paper clips, etc.). An extremely
valuable resource for teaching low-budget, hands-on
science with simple materials and without elaborate
facilities. The activities cover an amazingly wide range of
content.

The Flying Circus of Physics with Answers
By Jearl Walker
New York: John Wiley & Sons, 1977
www.wiley.com
An extensive, fascinating collection of problems and
questions about the real world. Answers and references
are given in separate sections at the end of the book.

College Physics (3d ed.)
By Jerry Wilson and Anthony Buffa
Englewood Cliffs, N.J.: Prentice Hall, 1997
Excellent addition to a reference library.

National Science Education Standards

In 1996, the National Research Council, a division of the National Academy of Sciences, published the *National Science Education Standards.* Copies are available from the National Academy Press (800-624-6242). The text of the *Standards* is available on the Internet at www.nas.edu. Shown on the next two pages is a list of the snacks in this book, and their relation to certain of the Content Standards for grades 5–8 and 9–12.

For the Physical Science Content Standards, we have shown a detailed correlation. For the Earth and Space Science, Science as Inquiry, and Science and Technology Standards, we have indicated whether the snack correlates to at least one of the more detailed standards. This treatment is not intended to be exhaustive. Some snacks may relate to other standards as well, and it may be possible to undertake any particular snack in such a way that it covers additional standards.

National Science Content Standards
(National Research Council)
Grades 5-8

	Physical Science		Earth and Space Science	Science as Inquiry	Science and Technology
	Motions and Forces	Transfer of Energy			
3-D Shadows				✔	
Bits and Bytes	✔		✔		
Circuit Workbench		✔		✔	✔
Diamagnetic Repulsion	✔	✔			
Film Can Racer	✔	✔			✔
Fractal Patterns			✔	✔	
Hoop Nightmares				✔	
Hydraulic Arm	✔	✔		✔	✔
Hyperbolic Slot					✔
Light Conversation		✔		✔	
Membrane Aerophone		✔		✔	✔
Modulated Coil		✔			✔
Modulated LED		✔			✔
Oil-Spot Photometer				✔	
Palm Pipes		✔		✔	
Periscope with a Twist				✔	✔
Personal Pinhole Theater				✔	✔
Perspective Window					✔
Pinhole Mirror			✔	✔	
Reverse Masks					
Saltwater Pentacell		✔		✔	✔
Sensitive Filament		✔			✔
Shadow Panel		✔			✔
Snip and Snap	✔	✔			
Soap Film in a Can				✔	
Sound Bite		✔			✔
Sprotating Cylinder				✔	
Square Wheels					✔
String Machine		✔		✔	✔
Stripped-Down Generator		✔			✔
Your Father's Nose					✔

Square Wheels • An Exploratorium Science Snackbook

National Science Content Standards
(National Research Council)
Grades 9-12

	Physical Science			Earth and Space Science	Science as Inquiry	Science and Technology
	Motions and Forces	Conservation of Energy and Increase in Disorder	Interactions of Energy and Matter			
3-D Shadows					✔	
Bits and Bytes	✔		✔	✔		
Circuit Workbench		✔	✔		✔	✔
Diamagnetic Repulsion	✔		✔			
Film Can Racer	✔	✔	✔			✔
Fractal Patterns		✔			✔	
Hoop Nightmares					✔	
Hydraulic Arm	✔	✔	✔		✔	✔
Hyperbolic Slot						✔
Light Conversation			✔		✔	
Membrane Aerophone			✔		✔	✔
Modulated Coil			✔			✔
Modulated Led			✔			✔
Oil-Spot Photometer		✔	✔		✔	
Palm Pipes			✔		✔	
Periscope with a Twist					✔	✔
Personal Pinhole Theater			✔		✔	✔
Perspective Window						✔
Pinhole Mirror					✔	
Reverse Masks						
Saltwater Pentacell		✔	✔		✔	✔
Sensitive Filament		✔	✔			✔
Shadow Panel		✔	✔			✔
Snip and Snap	✔	✔	✔			
Soap Film in a Can			✔		✔	
Sound Bite			✔			✔
Sprotating Cylinder			✔		✔	
Square Wheels						✔
String Machine			✔		✔	✔
Stripped-Down Generator		✔	✔			✔
Your Father's Nose						✔

About the Authors

Authors Don Rathjen and Paul Doherty experiment with light intensity while preparing the "Oil-Spot Photometer" snack.

About Don Rathjen

Don Rathjen has taught physics for more than thirty years in a variety of settings, including a small college in Liberia, West Africa, a Turkish high school, and public high schools in Pleasanton, California.

In 1985, Don began working with science teachers at the Exploratorium Teacher Institute. In this role, he was deeply involved in the development of the original *Exploratorium Science Snackbook,* published in 1991.

Don received a B.S. in Engineering from Stanford University and an M.A. in Science Education from Columbia University. His interest in using simple materials to teach physics and physical science was evident early in his career, when his own young children used to find their missing toys in Dad's physics lab. In 1983, Don began "Physics Day" at Great America, an amusement park in Santa Clara, California. This event has grown from an initial 90 students to more than 20,000 annually. Don is also the author of *Lego Crazy Action Contraptions,* published by Klutz Press.

Don and his wife, Ann, are proud to note that the next generation in their family is batting 6-for-6 in the teaching league—their three children and the children's three spouses are all middle school teachers! Don and Ann are thoroughly enjoying being grandparents—an experience that has included grandchildren building some of the snacks in this book with Grandpa!

About Paul Doherty

Paul Doherty is a physicist, teacher, author, and rock climber. Since 1986, he has been a vital part of the Exploratorium's Teacher Institute, delighting teachers and museum visitors with his exhibit-based explorations of scientific phenomena.

Paul received his Ph.D. in solid-state physics from the Massachusetts Institute of Technology in 1974. Over the years, he has shared his expertise in science and teaching through a variety of channels, including serving as a visiting

scientist at museums in Sweden and Brazil, teaching both through the Exploratorium and at the college level, and appearing as a guest on *Late Night with David Letterman.* He has written about science in magazines and books, including the *Klutz Book of Magnetic Magic, The Color of Nature,* and *Traces of Time.*

In 1996, Paul won the "Best Science Demonstrator" award at the World Congress of Museums in Helsinki for his virtuoso performance on the whirly—a corrugated plastic tube that sings when whirled. To find out more about Paul's adventures as a physicist, a teacher, and a rock climber, visit his Web site at www.exo.net/~pauld.

About the Exploratorium

The Exploratorium is a hands-on museum of science, art, and human perception dedicated to discovery. Founded in San Francisco in 1969 by the noted physicist and educator Frank Oppenheimer, the museum has grown over the years to become an internationally acclaimed science center. Its hundreds of interactive exhibits stimulate learning and richly illustrate scientific concepts and natural phenomena.

Since 1984, the Exploratorium's Teacher Institute has provided workshops for in-service science and math teachers of grades 6 to 12. The Institute offers a teacher-centered, learn-by-doing approach to science learning. Museum exhibits and classroom activities are used to explore science concepts and models using inquiry-based pedagogy. Each summer, the Teacher Institute offers several institutes—intensive, four-week, 100-hour programs—on a variety of science and math topics. Upon completion of the summer institutes, teachers become part of the Teacher Institute "family" of more than 2,000 alumni and are eligible to return for Saturday, after-school, and summer alumni workshops, as well as other special programs.

You can visit the Exploratorium online by pointing your browser to www.exploratorium.edu for more valuable science learning resources.

Acknowledgments

We want to thank Goéry Delacôte, Rob Semper, Dennis Bartels, and Linda Shore for providing the institutional backing necessary to transform this project from a vision into a reality.

Thanks to Tsing Bardin, who personally built and tested almost all the snacks in this book. Her comments from the perspectives of both teacher and physicist were invaluable. Thanks also to Chris Gibbons and his physics students at Foothill High School in Pleasanton, California, for an abundance of building, testing, and commenting.

Thanks to my wife, Ann, whose contributions to this book include our dining room table for protracted periods, our garage virtually permanently, enduring a Powerbook on vacations, and other virtues too numerous to mention.

Thanks also to our kids and grandkids for putting up with Dad/Grandpa and his toys and for the fun of building some of them together.

—Don Rathjen

Thanks to the staff of the Teacher Institute who not only inspired these snacks and suggested improvements, but also put up with the threat of avalanche from piles of snacks that build up to mountainous proportions around my desk.

—Paul Doherty

Contributors

This book has been several years in the making, and has involved many people in the process. The contributions of our colleagues in the Exploratorium Teacher Institute, other museum staff, and the many, many teachers who have been involved constitute the very being of this book.

Far more snacks were developed than it was possible to include in this volume, and many snacks have been through several generations of development by a variety of contributors. Transcending any particular snack, however, is the process itself. Participants had opportunities to bounce ideas off each other, gain new insight and inspiration from snacks initiated by their colleagues, respond to the challenge of determining a "better way to do that" as a particular snack evolved, exercise their creativity, build working prototypes, and come away with a host of new ideas. Ultimately, all snacks reflect a truly collegial effort, and in this spirit, we wish to thank the individuals listed below for their contributions. We also offer our sincere apologies to anyone who has been inadvertently omitted; if this has occurred, please let us know and we will attempt to rectify the error in subsequent printings if possible.

Vivian Altmann
Miguel Appleman
Chris Axley
Tsing Bardin
Pete Bentivegna
Laura Bergman
Charles Bissell
Heidi Black
Ian Bleakney
Pam Blinn
John Boccuzzi
Ron Bonstetter
Boston University
   Center for Polymer
   Studies
Tory Brady
Diane Bredt
Lynette Brown
Barbara Bureker
John Carmean
Yiu Man Chan
Rilla Chaney
Lourdes Chico
Christopher Chien
Craig Childress
Paul Ching
Edmund Chu
Coral Clark

Carri Cupp
Sheila Curtis
Frank Daar
James Debridge
Nancy Defensor
Pablo Dela Cruz
Richard Delwiche
Mish Denlinger
Gene Easter
Lewis Edinborough
Gabe Espinda
Kathleen Farrell
Neil Fetter
Ken Finn
Jennifer Fong
Art Fortgang
Robin Franklin
David Fryman
Curt Gabrielson
Galileo Circle
Keith Geller
Chris Gibbons
Wendy Goodfellow
Kevin Gortney
Rosa Haberfeld
Brett Hamilton
Russ Harding
Theresa Heckathorne

Sherry Hernandez-
   Woo
Glorianne Hirata
Stan Hitomi
Linda Hjelle
Fran Holland
John Holley
Thomas Humphrey
Tien Huynh-Dinh
Jane Jolly
Raymond Jones
Marco Jordan
Karen Kalumuck
Annamarie Karsant
Norman Keeve
Donald Kennedy
Burt Kessler
Eric Kielich
James Kliewer
Barry Kluger-Bell
Nancy Koch
John Lahr
Karen Laitinen
Lori Lambertson
Kendra Langer
William Lau
Carol Lee
Lowayna Lewis

Alice Liang
Don Linstad
Richard Lohman
Susie Loper
Robert Lovelace
Kaitlin Lowell
Regan Lum
Jamal Mpenda
Robert McClard
Karen Mendelow
Tim Merrill
Trish Mihalek
Lisa Miller
Richard Moore
Art Morrill
Dean Muller
Eric Muller
Fred Muller
Erainya Neirro
Linda Ng
Todd Nore
Heather O'Connor
Philip Ogata
Michael Palmer
Barbara Pinney
Sabrina Ramirez
Ann Rathjen
David Rathjen

Bill Reitz
Charles Reynes
Georgiana Rudge
Cora Salumbides
Linda Shore
Doug Spalding
Crans Squire
Fred Stein
Ted Stoeckley
Martin Stoye
Victoria Stroup
Candy Sykes
Modesto Tamez
Germaine Titus
Guillermo Trejo-Mejia
Adrian Van Allen
Julie Walker
Eric Watterud
Paula Weisman
Nancy Wilson
Yvonne Wong
Daryl Zapata
Allison Zbikowski
Barbara Ziegenhals
Philip Zike

Topic Index

General Index

Learning Tools from the Exploratorium

At the Exploratorium, we create environments and tools that awaken curiosity and encourage people to learn more about the world around them. If you have enjoyed *Square Wheels and Other Easy-to-Build, Hands-On Science Activities*, you will find these other Exploratorium publications of interest. You can order these products and learn about them online at: www.exploratorium.edu/store.

Science Activity Books for Kids and Families

The Brain Explorer: Puzzles, Riddles, Illusions, and other Mental Adventures. Pat Murphy, Ellen Klages, Pearl Tesler, Linda Shore, and the staff of the Exploratorium. Henry Holt and Company, 1999.

Explorabook: A Kid's Science Museum in a Book. John Cassidy and the Exploratorium. Klutz Press, 1991.

Glove Compartment Science: Experiments, Tricks, and Observations for the Backseat Scientist. John Cassidy, Paul Doherty, and Pat Murphy. Klutz Press, 1999.

The Science Explorer: An Exploratorium Science-At-Home Book. Pat Murphy, Ellen Klages, Linda Shore, and the Exploratorium. Henry Holt and Company, 1996.

The Science Explorer Out and About: An Exploratorium Science-At-Home Book. Pat Murphy, Ellen Klages, Linda Shore, and the Exploratorium. Henry Holt and Company, Inc., 1997.

Zap Science: A Kid's Science Museum in a Book. John Cassidy, Paul Doherty, Pat Murphy, and the Exploratorium. Klutz Press, 1997.

Books for Curious Adults

By Nature's Design. Text by Pat Murphy, photographs by William Neill. Chronicle Books, 1993.

The Garden Explored. Mia Amato and the Exploratorium. Henry Holt and Company, 1997.

The Inquisitive Cook. Anne Gardiner and Sue Wilson with the Exploratorium. Henry Holt and Company, Inc., 1998.

The Sporting Life. Susan Davis, Sally Stephens, and the Exploratorium. Henry Holt and Company, 1997.

Traces of Time: The Beauty of Change in Nature. Pat Murphy and Paul Doherty, photographs by William Neill. Chronicle Books, 2000.

Watching Weather. Tom Murphree and Mary Miller with the Exploratorium. Henry Holt and Company, Inc., 1998.

Building Science Exhibits on a Shoestring: The Exploratorium Science Snackbook Series.

The Cheshire Cat and Other Eye-Popping Experiments on How We See the World. Paul Doherty, Don Rathjen, and the Exploratorium Teacher Institute. John Wiley & Sons, 1995.

The Cool Hot Rod and Other Electrifying Experiments on Energy and Matter. Paul Doherty, Don Rathjen, and the Exploratorium Teacher Institute. John Wiley & Sons, 1996.

The Spinning Blackboard and Other Dynamic Experiments on Force and Motion. Paul Doherty, Don Rathjen, and the Exploratorium Teacher Institute. John Wiley & Sons, 1996.

Science in the Classroom: Books for Students and Teachers

The Exploratorium Guide to Scale and Structure: Activities for the Elementary Classroom. Barry Kluger-Bell and the School in the Exploratorium. Heinemann, 1995.

Human Body Explorations: Hands-On Investigations of What Makes Us Tick. Karen Kalumuck and the Exploratorium Teacher Institute. Kendall/Hunt Publishing Co., 2000.

Exhibits and Exhibit Building: The Exploratorium Cookbook Series

Exploratorium Cookbook I. Raymond Bruman and the Staff of the Exploratorium. The Exploratorium, 1991.

Exploratorium Cookbook II. Ron Hipschman and the Staff of the Exploratorium. The Exploratorium, 1983.

Exploratorium Cookbook III. Ron Hipschman and the Staff of the Exploratorium. The Exploratorium, 1987.

Charts and Posters

The Ball Makes the Game—A Poster with Activity Guide.

Cycles Are Everywhere—Poster and Activity Guide.

Electromagnetic Spectrum Chart—Poster.

Language Families of the World—Poster and Activity Guide.

Old Woman or Young Girl?—Optical Illusion Poster.

Portrait Goblets—Optical Illusion Poster.

Sound Spectrum Chart—Poster and Activity Guide.

A Story of Letters: The Evolution of the Modern Roman Alphabet—Poster and Activity Guide.